Atlantis, Found?

An Investigation into Ancient Accounts,
Bathymetry and Climatology

Second Edition

2018

Jonathan Northcote

Copyright

Publisher: Jonathan Northcote (Self-published)
Email: jonathan.northcote@gmail.com
Second Edition

First published: Ebook 2018

Dedication

This is dedicated to my family.
Thank you for your support and assistance.

Table of Contents

Preface

After publishing my first edition of this book (under the title 16.484°W 58.521°N Atlantis, Found?) I sent a copy to Tony O'Connell, the webmaster of Atlantipedia (www.Atlatipedia.ie), an extremely useful, extensive and encyclopaedic site on all things relating to Atlantis.

Tony O'Connell, although he did not agree with my conclusions, was kind enough to comment on my book. His comments, both complementary and critical, are published on his website. He said that there is *"no doubting the quality of Northcote's research"* and that I offer *"spirited support for* [Hatton Rockall being Atlantis] *using a mass of geological data and underwater topography"*. He, however, specifically questioned, and disagreed with, certain aspects relating to *"words and phrases such as, continent, Pillars of Herakles, Atlantic, greater than Libya and Asia combined and elephants."*

In this edition I have made some substantial changes; I have changed the layout of the book, attempted to make my writing less formal or academic, included, (where I felt I could do better than I had done in the original edition) responses to the issues raised by Tony O'Connell, included certain new information and references and corrected certain typographical errors.

I have, both in this edition and in the original edition, referred to a number of Wikipedia articles. I recognise that these are not always regarded as good authority - at least in academic circles. Because of this, my references to Wikipedia have, generally, been with regard to minor, or side, issues and matters which I suspect are uncontentious. That said, I have found such articles to provide useful summaries and good starting points. Wherever possible, and certainly for the important aspects of my research, I have referred to original works.

Thank you to all who bought my original edition. This edition will, I hope, provide answers to the queries raised by Tony O'Connell and any that other readers may have.

Preface to the First Edition

IT MAY LEGITIMATELY BE ASKED of me 'What prompted you, a legal practitioner, to write this book?'

The answer is that during 2011 I read Otto Muck's book *The Secret of Atlantis*.[1] Shortly thereafter I happened to see, over the shoulder of someone who was looking at Google Earth, the image of a submerged area, in the North Atlantic, which appeared to have a central plain with, in turn, a centrally placed mount/rise. The area, I have now come to learn, is the Hatton Rockall plateau and the rise in the middle of its central plain is Mammal Seamount.

Having been somewhat dissatisfied with the argument and conclusions in *The Secret of Atlantis* I decided to find out more about the area and whether there were any further similarities. The more I investigated, the more similarities I found between Hatton Rockall and the descriptions given, by Plato, of Atlantis. My curiosity having been piqued by these physical similarities, I decided to investigate still further and, ultimately and as a consequence, decided to write this book.

I do not propose that I supply all the answers – on the contrary there are many questions unanswered and avenues for further investigation which, I hope, might warrant further investigation by others more knowledgeable in the relevant fields.

I must also, in conclusion, acknowledge that I have discovered, in the course of researching the topic, that I am not the only person to suggest that Hatton Rockall was Atlantis. There are, to my knowledge, at least three other individuals[2] who have proposed this.

What, I believe, in particular distinguishes[3] my work from those of these other authors is the extent of the research, or quoted research, incorporated into my work (see the reference sections at the end) and that I attempt to supply an explanation for the sinking.

SECTION ONE: INTRODUCTION

1.1   Introduction

The story of the ancient island state of Atlantis[4] and how it is said to have sunk beneath the waves is known by many people and has generated much interest over the millennia since Plato first wrote about it in about 360 BCE.

Although Plato appears to be the first, and only, person to have written about the fabled isle of Atlantis there are many hints of an idyllic island in the Western Ocean. Probably the oldest Western reference is by Homer, in his *Odyssey*, to the island of Ogygia, the home of the nymph Calypso. Calypso was a daughter of the god Atlas and her other, patronymic name, was 'Atlantis'.

There have been many attempts to substantiate various aspects of the Atlantis story but the location of Atlantis has yet to be determined (or its very existence accepted) and it still attracts attention and debate. Studies in scientific journals[5] have also been included in these attempts. (Similar scientific attention has also been given to other myths and legends.[6])

Notwithstanding these many attempts, the existence of Atlantis remains outside of generally accepted history. But, accepted history must and should be (along with other accepted ideas) reconsidered and tested from time to time. As the theoretical physicist Carlo Rovelli stated – *"Science is not about certainty. Science is about finding the most reliable way of thinking, at the present level of knowledge. Science is extremely reliable; it's not certain. In fact, not only it's not certain, but it's the lack of certainty that grounds it. Scientific ideas are credible not because they are sure, but because they are the ones that have survived all the possible past critiques, and they are the most credible because they were put on the table for everybody's criticism."*[7]

It is in the above light that I advance my hypothesis.

1.2 The approach adopted and the extraction of essential features.

William H. Babcock, in his book *Legendary Islands of the Atlantic – A Study In Medieval Geography*[8] commented: "*It is evident that the Atlantis tale must be treated either as mainly historical, with presumably some distortions and exaggerations, or as fiction necessarily based in some measure (like all else of its kind) on living or antiquated facts. Certainly no one will go the length of accepting it as wholly true as it stands. But, even eliminating all reference to the god Poseidon and his plentiful demigod progeny, we are left with divers*[e] *essential features*".

Similarly, Professor Duane Roller, in his book *Through the Pillars of Herakles, Greco-Roman exploration of the Atlantic*[9] (where he discusses the relevance of Iceland, the Arctic and the world beyond the Mediterranean Sea to Greco-Roman culture) comments, in the Introduction, that "… *there is a whole genre of fantasy literature about travel to remote places, beginning with the Odyssey. These accounts often mixed in actual data, and so can be valuable sources that may be difficult to untangle. … Fantasy writers, even today, are always skilled in mixing the real and the unreal, siting their tales just beyond the limits of human knowledge…*"[10]

I shall focus on the "*divers*[e] *essential features*" of the Atlantis story in an attempt to untangle the facts from fiction.

Being a legal practitioner by profession, I shall approach the story from a quasi-legal stance; comparing a set of allegations with known facts. I shall examine the description given by Plato of the island of Atlantis, extract the diverse "*essential features*" from the description given by him, assume these to be the core 'allegations' describing Atlantis and then view these 'allegations' in the light of known facts, modern knowledge and theories. This will be done with the intention, and hope, of persuading you that Plato, when describing Atlantis, was describing a real place and, specifically, what is now known as the Rockall Plateau. It would, of course, be impossible to truly verify, beyond all doubt, Atlantis' existence and location but, through this book, I seek to persuade you that it is probable[11] that Plato was describing a real place and that is the Rockall Plateau.

The Rockall Plateau (which I shall refer to as simply Hatton Rockall as it encompasses both the Hatton and Rockall highs) is a large submerged area in the North Atlantic to the west of Ireland and the United Kingdom.

I shall not, for the most part, deal with the descriptions given of the Atlantean civilisation and its general trappings - I take the view that Plato, being a philosopher, was probably using the story of the Atlantean civilisation[12] as a 'vehicle' to emphasise a philosophical point.[13] Plato, in addition, explicitly acknowledges that certain of the original information has been changed by (at least) giving the Greek versions of names.[14] It is even possible, on this basis, that 'Atlantis' was not the original, or native, name of the island but merely the name given to it by Plato. I suggest that the name *Atlantis* was probably chosen by Plato, using the patronymic name of Atlas' daughter Calypso (this will be elaborated upon in section 3.3), as a name which his audience would have known and understood.

For these reasons I shall, in extracting the details from Plato's description of Atlantis, attach more weight to certain aspects of the description and less weight to others.

Estimates of distance and of time are often regarded by our courts with a degree of circumspection and as often being unreliable – unless supported by other evidence. I shall treat with some circumspection the statements regarding to the precise size of Atlantis and the stated date of the disappearance of Atlantis (9 000 years is, by most human measures of time, an immense time period). However, as will be seen in Section 11, there were cataclysmic events which took place around the time that Atlantis is stated to have been submerged. The physical features of the island are, in my opinion, less likely to have been made up or greatly exaggerated and so shall be accepted more or less at face value.

For those who may doubt that it was possible for people to have populated an island that long ago, it should be borne in mind that Australia (which has always been surrounded by ocean) was populated by the Aborigines as early as about 65 000 years ago.[15,16] The amount of open sea between Hatton Rockall and the mainland at that time was not as large as it is now and the weather around Hatton Rockall was not (as will be seen in section 8.3) as inclement as current times lead us to believe.

1.3   The structure of this book

There are five broad sub-divisions in the book.
> ➤ The first sub-division covers the issues of

> > ➤ the possibility of facts being conveyed with reasonable accuracy down the millennia (and thus, how plausible it is, that the description of Atlantis could likewise have been accurately conveyed) (Section 2) and

> > ➤ a consideration of idyllic islands in the western ocean referred to in, for the most part, Greek mythology (Section 3),

> ➤ The second deals with Plato, his description of Atlantis, the issues (mentioned in the Preface) concerning the words and phrases 'Continent', 'Pillars of Heracles', 'Atlantic' and 'greater than Libya and Asia combined' and where the island of Atlantis was situated (Sections 4, 5 and 6),

> ➤ The third is a comparison of Plato's description of the "*essential features*" (particularly the physical geography and climate) of Atlantis with known facts concerning the Hatton Rockall area, as well as considerations of the massive flooding, earthquakes and volcanism around the stated time of the destruction of Atlantis, what, or where, Gadeira/Gades might be and certain other circumstantial evidence. (Sections 7 to 11)

> ➤ Having, hopefully, persuaded you that the congruencies between Atlantis and Hatton Rockall go beyond mere coincidence (section 12 contains a summary of these) I shall then move on to the fourth sub-division which is more speculative. In Section 13 I shall consider various other factors possibly explaining or relating to the cause and mechanism of Atlantis' sinking and how long it took to sink and

> ➤ The fifth is an appendix with depth profile diagrams of Hatton Rockall showing the contours and shape of Hatton Rockall.

I am not an expert on any of the aspects or fields of knowledge to which I shall refer but have endeavoured, firstly, to collect information from across many disciplines, and, then, to present the information as a coherent body of evidence in support of the proposition that the island of Atlantis was, what I refer to as, Hatton Rockall.

Ultimately I hope that you will, after considering all the evidence put forward, be persuaded that when Plato described the island of Atlantis he was in fact describing, what is known as, the Rockall Plateau, or Hatton Rockall area.

SECTION TWO:
CAN INFORMATION BE PASSED DOWN ORALLY, AND ACCURATELY, OVER MILLENNIA?

2.1   Introduction

Plato is the not the only person to have described an idyllic island in the western ocean. There are many references to idyllic, or 'Blest', isles in the extreme west (See section 3) but none of these supply the extent of detail given by Plato, nor do they name the island 'Atlantis'.

Before we consider Plato's description there is one question which even the most ardent believer in Plato's Atlantis must surely ask and that is 'How is it possible that the story of Atlantis could have been passed down through nine thousand years without being written down?' And passed down accurately!?

In order to answer these questions it is necessary to consider

➤ whether there are other records of memories, or stories, of historical events having been passed down over extended periods; even millennia,

➤ the, stated, original source of Plato's information and how he, Plato, could have obtained the information.

2.2   Transmission of information

What supports the possibility that the information might have been conveyed, relatively accurately, down the millennia and centuries?

There are, with certain degrees of overlap, two aspects to considering this issue, namely

> is it plausible that such information could have been passed down over millennia and

> why should this information have been retained over the millennia, by the Egyptians, and centuries, by Solon's descendants?

I shall first consider the second of the above issues and then deal with the first.

2.3   Why should the information have been retained?

Most reasonable people will, in my view, go to the most effort to store or retain information if they believe that information is both important and trustworthy.[17]

This being so, one needs to assess the importance of the story of the Atlantis flood and its trustworthiness, or likely veracity. If the Egyptians, Solon, and then his descendants, considered the story to be important and true it is far more plausible that the story could have been passed down from generation to generation.

2.3.1   The importance of the events described

At the start of the Holocene, about 10 000 BCE, the earth (or, at least, the northern hemisphere) was undergoing rapid and fundamental changes. It was experiencing unprecedented and extremely powerful earthquakes, floods and volcanism (as I will show in the following sections) and significant changes in climate.[18] These occurred, in all likelihood, across the entire northern hemisphere and, probably, caused a significant reduction in the human population. That there was a significant reduction in the human population is borne out by a recent study[19] which found that there has, within the last ten thousand years (i.e. since about 8 000 BCE), been a genetic bottleneck[20,21] of Y chromosome diversity (and that this is linked to changes in culture). *Karmin et al's*. article shows (Figure S4B in the Supplementary Figures document) that the effective population sizes of all Y chromosome populations were, as at 10 000 BCE, approximately the same. However shortly thereafter the effective Y chromosome population size in Africa started to drop. Significant drops in the effective Y chromosome population in virtually all populations commenced by about 8 000 BCE. There was also a genetic bottleneck specific to the Finnish population between 10 000 and 20 000 years ago.[22,23] These bottlenecks were, probably, caused by the dramatic and unprecedented changes in the environment.

People would have been subject to threats to their health and safety on scales still unknown to us today. Their very survival was threatened as a result of the cataclysmic floods, earthquakes, volcanism, weather and climatic extremes and the concomitant impacts of these on food yields, fresh water and infectious diseases. These would all, in turn, also have caused conflict and displacement of peoples.[24,25]

It is reasonable to accept that a memory of such unprecedented and traumatic events would have been regarded as important to be stored and passed on to further generations. These memories would, without the benefit of writing, have been told and retold amongst survivors and passed down in the form of sagas and myths. That memory would also, accepting that the destruction of Atlantis was part of these events, have been passed down over the generations.

2.3.2   *The trustworthiness of the story and its information*

The issue of the perceived trustworthiness of the information to be retained, or passed on, may be phrased as 'Does the person receiving the information believe that the memories, or information, could have been retained and passed down accurately over extended periods?'

Let me start by acknowledging that it appears, at first glance, extremely implausible that the description of an event and place (no matter how significant) could be conveyed, firstly, over millennia and, secondly, with any degree of accuracy. This is so, particularly, for us who live in a culture in which writing is the primary and paramount manner for the transfer and recording of knowledge. Yet even in the most literate and Western of societies we retain sayings, customs and beliefs which originated hundreds, or even thousands, of years ago. In Christianity the Old Testament would, before it was canonised, have been passed down orally.

We also often teach our children nursery rhymes, the origins of which have been lost in the mists of time but we still pass them on believing them to be good or necessary.

Even in certain written cultures today there are traditions of being able to learn, and repeat, large tracts of knowledge. There are people today, referred to by Muslims as '*Hafiz*', who have completely memorised the Qur'an[26] – which is divided into 114 Surahs, or chapters, containing 6 236 verses and comprising some 80 000 words. The memorisation of the Qur'an was, at least in the past, considered more secure than having it written down — a manuscript could easily be destroyed, but if the Qur'an was memorised by many *huffaz* it would never be lost.[27] Muslims believe that the Qur'an was revealed by God/Allah to Muhammad gradually over a period of approximately 23 years, beginning in 609 CE, when Muhammad was 40, and concluding in 632 CE, the year of his death. The Qur'an did not exist in book form at the time of Muhammad's death[28] and, accordingly, it must be accepted that the oral recitation of the Qur'an, from memory, has been taught and is ongoing, without any change (or significant change) to the Qur'an, for nearly 1 400 years.

There are, in addition, many other spiritual or religious traditions amongst peoples who were, until comparatively recently, non-literate, which can be reliably dated to centuries or even millennia in the past - as we shall see in the next sub-section.

Professor Nunn, in his book *The Edge of Memory: The Geology of Folk Tales and Climate Change*,[29] cautions that we should not measure the cognitive abilities of non-literate people in a non-literate world against our own, literate, abilities and expectations in a literate world. Put more colloquially, we must ensure that we compare apples with apples and, possibly, do not patronise our forebears.

2.4 Plausibility of the reliable transmission of such information

In support of the proposition that Plato could have relied upon an extremely ancient myth I shall refer to certain other myths which evidence the underlying 'truths' and the age of these myths.[30, 31]

I do not use the term 'myth'[32] in any pejorative sense, but, on the contrary, will throughout attempt to follow a 'geomythological approach'. The term 'geomythology' was coined by Dorothy B. Vitaliano during 1967. She defined a geomythologist as one who *"seeks to find the real geologic event underlying a myth or legend to which it has given rise; thus he helps convert mythology back into history."*[33,34]

Roger Echo-Hawk proposes[35] that there are three tests, or a threefold test, to be applied to when considering the credibility, trustworthiness and importance of oral traditions and myths. The first is that it should, at best, exhibit only vague chronological indicators in its relationship to historical events mentioned in other traditions. Secondly, it should be presented as a story about events presumed to be historical and thirdly it must be supported or verified through evidence gathered from non-verbal sources, such as archaeological data, written records or other accepted sources of evidence.

The myths considered below meet these requirements.

2.4.1   'Recent' myths

It is not uncommon to find 'stories' recording factual incidents which are hundreds of years old. By way of example; Bryant *et al.*[36] consider some Australian Aboriginal and Maori myths that could describe cosmogenic impacts in the ocean and the resultant mega-tsunamis.

There is a crater, about 20 km across on the continental shelf about 250 km south of New Zealand and under about 300 m of water, Based upon the lack of, or minimal, sediment in the crater they conclude that it must be less than 1 000 years old. After considering four different lines of evidence, they also concluded that there was probably a comet impact during the 15th century and that the crater is the impact site. Such an impact would, on all reasonable probabilities, have caused a massive tsunami. This deduction appears to be corroborated by certain myths, and other evidence, from the Kimberley area of Australia which seem to indicate that there was, in fact, a mega-tsunami off the north-west coast of Australia around 1690 CE/AD.

Similarly, in the article '*The fall of Phaethon: a Greco-Roman geomyth preserves the memory of a meteorite impact in Bavaria (south-east Germany)*'[37] the authors, B. Rappenglück *et al.*, posit that the myth of Phaethon records the fall of a meteorite and its impact which took place between about 2 200 to 800 BCE at Chiemgau, Bavaria.

2.4.2 *Myths having an ancient basis*

In addition to the myths such as the above recording relatively recent events there are also records, from a number of continents, of myths which indicate a recording of events which took place many thousands of years prior to the present.

It is to these myths that I wish to draw attention – given that the story told by Plato was (at the time that he told it), ostensibly, 9 000 years old.

2.4.2.1 North America

Vitaliano tells[38] that, according to the myths of the Klamath Indians of America, the Crater Lake of Mount Mazama was formed as a result of the fight between the chiefs of the above and below worlds who threw rocks and flames at each other. The fight only ended when Mount Mazama, on which the chief of the underworld had been standing, collapsed under him and sent him back to his underworld domain. The hole left by the collapse of the mountain then filled up and became Crater Lake.

This 'story' is corroborated by scientific knowledge. Crater Lake was, according to geologists, created by the collapse of a volcano which subsequently filled with water. It is stated, in Wikipedia, that:

"Mount Mazama was a stratovolcano in the Oregon segment of the Cascade Volcanic Arc and the Cascade Range located in the United States. The volcano's collapsed caldera holds Crater Lake, … . When it last erupted, the eruption was 42 times greater than the eruption of Mount St. Helens in 1980.

Mazama's summit was destroyed by a volcanic eruption that occurred around 5,677 (± 150) BC.[39] The eruption reduced Mazama's approximate 12 000-foot (3 700 m) height by around a mile (1 600 m). Much of the volcano fell into the volcano's partially emptied neck and magma chamber. "[40]

Douglas Deur, in his 2002 article *A Most Sacred Place: The Significance of Crater Lake among the Indians of Southern Oregon,*[41] considers the similarities between the descriptions given in the myth and a geological description of the eruption. He concludes that *"the sequence of events recounted by Klamath elders suggests that their oral history was informed by first-hand accounts of the same sequence of eruptions described by modern geologists"*

The memory of this massive eruption had therefore been passed down, in oral myth form, for nearly 8 000 years.

2.4.2.2   Australia

Robert Dixon records that he was informed of an Australian Aboriginal myth (or 'Dreamtime' story) about the formation of Lake Euramoo. The lake is a shallow volcanic crater lake in North Queensland, Australia. According to the geological information available, the lake was formed about 10 000 years ago by two massive explosions from groundwater super-heating. The description given in the 'story' of the formation of the lake, and two other companion lakes,[42] has been said to be a plausible and a surprisingly accurate oral account of volcanic eruptions or explosions in the area. The man who conveyed the story to Dixon also commented that, at the time of the events, the vegetation had been different from that found at the lake at the time. This was simply noted by Dixon but he could do nothing with this. Some four years after the telling of the story, another scientist showed, from studying the sediments of the lake, that the predominant rain forest (which had existed prior to clearing) in the area is only about 7 600 years old.[43,44] This being so, it must be accepted that this 'story' was at least 7 600 years old and probably as old as 10 000 years.

Nunn and Reid in their article *Aboriginal Memories of Inundation of the Australian Coast Dating from More than 7000 Years Ago*[45] discuss various Aboriginal oral traditions of coastal inundation. They conclude, based on the current depth of various places described in the 'stories', that most of the stories refer to a period of postglacial sea-level rise which occurred more than 7 000 years ago and that some of these 'stories' may be up to ~ 13 000 years old.

2.4.2.3   Europe

In Scandinavia and Iceland there appears to be a continuous cultural evolution from the Bronze Age (and possibly even the, earlier, late Stone Age) through to the Vikings. The religion that prevailed before the Vikings, and before Christianity, was the Asa faith. This faith, or aspects thereof, was recorded in writing in the 13th century CE in Iceland in the

Edda.[46] There are two forms of the Edda[47] – both of which record many myths dating back unknown hundreds or even thousands of years. Some of these myths appear as if founded in geological events however, until comparatively recently, it was generally accepted that Scandinavia had been largely free of strong seismic activity. It has become clear in the last, approximately, thirty years that areas of Scandinavia were subjected to strong seismic activity at the time of deglaciation, about 12 000 years ago (as is referred to in Section 11).

It has been suggested by Morner[48] that these events, at least in the late Holocene, have influenced Norse mythology and possibly ancient place names. He refers[49] to a number of place names that could be related to earthquakes and the noises that occur with earthquakes. Amongst others he refers to Lake Dunkern (at which an earthquake occurred at about 6 000 BCE) whose name refers to deep noise and to Lake Hjälmaren (which falls within an area traversed by faults which were active in post-glacial times) and whose name conveys a sense of noisiness. Also of particular interest is the fact that the name 'Pärvie' (to which I refer in section 11.3.3) also refers to a noise from underground but, this time, in the language of the Laps. He suggests, in relation to both the Pärvie fault and Lake Dunkern, that the seismic events which took place at these locations seem to have been far too early to have affected the place name. As a consequence he suggests that there may have been subsequent activities which gave rise to the names. There appear, however, to be no record of such later significant events.

Morner's suggestion (i.e. that these event took place too early to have affected place names) appears to be made purely on the basis that it is unsupported and goes contrary to accepted (Western) history. I suggest, given the above Aboriginal and American Indian myths, that it is possible that the seismic events at the beginning of the Holocene could, in fact, have affected the naming of the places and that Morner was possibly overly hasty in rejecting these as possible bases for ancient place names.

In chapter five of his book *The Edge of Memory: The Geology of Folk Tales and Climate Change*, Nunn considers various myths of inundation from Brittany, Cornwall and Wales. His conclusion is that, taking into account scientific records of sea level changes in these areas, these myths could describe events that took place more than 10 000 years ago.

Finally, as regards European myths, it has been suggested[50] that that the impact of the Kaali meteorite in about 5 000 BCE,[51] might be recorded in, or have had effects upon, certain Estonian and Finnish mythology. If this is so then this myth has endured for about 7 000 years.

2.4.2.4   Asia

The area around Mount Pinatubo on Luzon Island in the Philippines is inhabited by, amongst others, an aboriginal group called the 'Ayta'.[52] They have inhabited the area for unknown ages.[53,54]

Although, prior to 1919, Mount Pinatubo was only recognised by a few geologists as being a volcano, the name 'Pinatubo' is formed from the root '-tubo' which, in both the Filipino (Tagalog) and Ayta languages, relates to anything concerned with growth. Together with the prefix 'pina-' it creates a word meaning "made to grow" or "allowed to grow".[55] This has led volcanologists to speculate that the Aytas witnessed a previous post-eruption growth of a new peak.

One does not know how far back the origins of the knowledge of the growth of the volcano extend and whether it results from only one eruption or many. It has been hypothesized[56] that the 'modern' series of six prehistoric eruptions began around 33 000 BCE with the last two of the previous eruptions[57] having taken place about 3 900 to 3 300 and 800 to 500 years ago respectively.

2.4.3   Conclusion

It therefore appears possible that information, if sufficiently dramatic and important, could have been passed down accurately over millennia in the form of myths.

2.5   The Atlantis story

2.5.1   The original source of the Atlantis story

According to Plato, the story of Atlantis was obtained from Solon who, in turn, obtained the information from a priest at Sais in Egypt.

Solon was an Athenian statesman, lawmaker, and poet who is remembered for his efforts to legislate against political, economic, and moral decline in archaic Athens. He lived from about 638 BCE to 558 BCE. He was granted, as Athenian ruler and lawmaker, extra-ordinary powers to make laws and then, with those powers, passed a number of reforming laws. After making those laws (and to prevent him from being prevailed upon to change them) he left Athens,[58] reputedly going to Egypt. He was, according to the *Timaeus*, well received there[59] and, whilst there, went to the temple at Sais,[60] an ancient Egyptian town in the Western Nile Delta.

Sais was, at the time, the capital of the ruling dynasty of Egypt. It was also the centre for the worship of the goddess Neith.[61,62] Neith[63] is one of the earliest and most revered goddesses in Egypt, possibly going back to 'primeval times'.[64] 'Concrete' records, in the form of a drawing on an ivory label, go back to about 3 100 BCE. Amongst her other attributes, Neith was the goddess of floods and even the personification of the primordial waters of creation.[65]

If there was any group of priests, or religion, whose records, or histories, should include stories of a great flood it would surely be the goddess of floods. This being accepted, it would then be reasonable to in turn accept that the priests of Neith at Sais would have been likely to have had stories and histories of the 'primeval' flood and its effects.

The question still remains however as to why, apparently, there is no other record of the Atlantis story. (Although it appears to be accepted that there is no other record, see the discussion in section 3.2 of the Edfu texts and the story of the destruction of the primeval island.)

The answer to the above question may be that the information which was given to Solon was, apparently, not common knowledge. Solon had, in essence, to trick the priest[66] into divulging the information. In the *Timaeus*[67] Plato wrote "*Hither came Solon, and was received with honour; and here he first learnt, by conversing with the Egyptian priests, how ignorant he and his countrymen were of antiquity. Perceiving this, <u>and with the view of eliciting information from them,</u> [my emphasis] he told them the tales of Phoroneus and Niobe, and also of Deucalion and Pyrrha, and he endeavoured to count the generations which had since passed. Thereupon an aged priest said to him: 'O Solon, Solon, you Hellenes are ever young, and there is no old* *man who is a Hellene.' 'What do you mean?' he asked. 'In mind,' replied the priest, 'I mean to say that you are children; there is no opinion or tradition of knowledge among you which is white with age; and I will tell you why. Like the rest of mankind you have suffered from convulsions of nature, which are chiefly brought about by the two great agencies of fire and water. … The memorials which your own and other nations have once had of the famous actions of mankind perish in the waters at certain periods; and the rude survivors in the mountains begin again, knowing nothing of the world before the flood. But in Egypt the traditions of our own and other lands are by us registered for ever in our temples."*

It can be seen, from the above, that Solon only obtained the information by subterfuge. The information appears to have been secret or restricted. Otherwise, Solon would not have felt obliged to trick the priest into disclosing the information. It is also plausible that, because the priest was dealing with the respected leader of Athens, he felt willing to disclose this restricted information to him.

Another possible answer to the above question may lie in the fact the priest would not have referred to the island as 'Atlantis' and if people have looked for an Egyptian myth about Atlantis they would not have found one. The word 'Atlantis' is derived from the name of the Greek god Atlas. Plato admits that the names used in the story have been changed to Greek / Hellenic names. He writes, in the *Critias*[68]: *"I ought to warn you, that you must not be surprised if you should perhaps hear Hellenic names given to foreigners. I will tell you the reason of this: Solon, who was intending to use the tale for his poem, enquired into the meaning of the names, and found that the early Egyptians in writing them down had translated them into their own language, and he recovered the meaning of the several names and when copying them out again translated them into our language."*

2.5.2   *How Plato obtained this information*

Plato (as will be elaborated upon in Section 4) lived from about 428 to 348/347 BCE and was, possibly the greatest philosopher in the history of western civilisation.

He does not give any explanation of how he obtained the information about Atlantis but both Plato and the eponymous Critias were related to Solon, albeit that they were separated by centuries from him. Whether

Critias, in fact, related the story to Plato, whether the story was a family 'tradition' passed down or whether Solon had prepared a draft of his intended poem which had been passed down the family one cannot know. It is, also, possible that Plato himself might have travelled to Egypt.[69] Strabo claims[70] to have been shown where Plato lived when he visited Heliopolis in Egypt. Either scenario is possible but, in any event, specifically how Plato obtained the information is of limited importance.

What is, ultimately, important is not from where Plato obtained the information but whether the information which he did obtain could, plausibly, have been retained and passed down accurately over such an extended period of time.

2.6   Conclusion

Considering the myths referred to above, the fact that they have a factual basis and the time periods over which those myths have continued to convey memories of dramatic events, it is apparent that memories of cataclysmic events which had taken place many thousands of years ago may (particularly in pre-literate societies) be retained, and passed down, in cultural memories and myths.[71]

That being so, there is a reasonable prospect that Plato, when telling the story of Atlantis, was recalling ancient events, the stories of which had been passed down through the millennia.

SECTION THREE:
OTHER MYTHS AND LEGENDS OF IDYLLIC WESTERN ISLES

3.1   Introduction

There are many references, in ancient Greek and European historical and mythological records, to islands to the west in the Atlantic. They generally describe the island (or islands) as very fertile and climatically idyllic; very similar to the way in which Plato described Atlantis. These disparate descriptions might, taken together, corroborate Plato's story or, at least, hint at what remained of the Atlantean island after its flooding.

Before dealing with the ancient European descriptions of mythical islands there is one other possibly more ancient, but less certain, Egyptian description of an island which was flooded in primeval times and was then re-inhabited.

3.2   The Egyptian primeval island

In the later stages of Egyptian history there were three predominant religious or cosmological systems; namely those based at Memphis, Heliopolis and Hermopolis. In addition to these were significant sacred sites found at Edfu and at Sais. Although the various cosmological systems were ostensibly separate, they borrowed from each other and from earlier 'classic' ages, texts and cosmological systems.[72,73] They appeared to be interdependent to a certain extent.

The hieroglyphic texts at the temple at Edfu were studied, analysed and interpreted by E AE Reymond. In her article *The Shebtiw in the temple at Edfu,*[74] she states that the record at Edfu clearly sets out that the tradition of the creator gods, or *Shebtiw,* was based upon another much earlier religious basis, possibly obtained from Heracleopolis whose own tradition was based upon a tradition also brought from another place. The god whose worship was centred on Heracleopolis was Heryshaf, a creator and fertility god who was reputedly born from the primordial waters. [75] (He was also associated in Greek mythology with Heracles.)

Reymond subsequently published a book, based on her studies and doctoral thesis, entitled *The Mythical Origin of the Egyptian Temple*.[76] In this book she reiterates that the beliefs recorded at Edfu were based upon mythological traditions and beliefs which extended far beyond historical times.[77]

According to Reymond the Edfu texts begin with a description of the primeval island on which the gods were supposed to have lived.[78] The island was formed in the middle of the primeval ocean, Nun. It was the centre of a two stage creation of the world. The first was the earlier mythical physical creation of the island of the gods. The second was a recreation after the destruction of the 'first' world and was regarded as being more historical,

During the initial creation the formation of the primeval island was by the ancient gods. There was no further significant physical development during the first stage of creation, save that it appears to have been extended or enlarged by the formation of a mound.[79] The period of existence of the, generally nameless, gods during this initial period was understood to be limited. They existed for an unknown, but finite, period.[80] At the end of the initial period the 'world' was destroyed.

There is no clear and unambiguous statement as to how the first creation was destroyed. It appears to have been through some disaster during which darkness fell on the land,[81] there was a storm and an attack by an enemy pictured as a snake.[82] The destruction wrought was so violent that the primeval land was submerged,[83] destroyed[84] and the inhabitants killed.[85]

The second generation of inhabitants on the island (the *Shebtiw*) came from the water, or sea.[86] They arrived after a period of time and once the darkness had dissipated.[87] On their arrival at the island the new inhabitants found the surface of the water, and the land as well, covered with reeds.[88] The reeds were subsequently removed[89] and the *Shebtiw* proceeded, together with others, with the re-creation of the world.

This shares marked similarities with Plato's description of the destruction of Atlantis, save that it allows for the resettlement and redevelopment of the island.

It has been queried why there are no records of the story of Atlantis in Egypt. Given this translation by Reymond, the recognition that the story came from ancient sources and that there was borrowing (probably with name changes and emphases) from other Egyptian doctrines, it appears that Egypt did have a record of a society (of gods), which existed, in primeval times, on an island in the sea, which island was destroyed (during a period of great destruction and upheaval) by being submerged. There is a difference however in that the island was subsequently resettled.[90]

Reymond's translations accord with my proposition (as set out in sections 9 and 10) that the island of Atlantis was probably flooded and inundated but did not, at least initially, submerge beneath the sea.

I shall now move on to deal with the European myths about islands in the far western ocean.

3.3 Ogygia

3.3.1 *Homer*

Homer is revered as the greatest of the ancient Greek epic poets. It is not known when he lived[91] but modern research places him in the 7th or 8th centuries BCE (about 4 centuries earlier than Plato) although other ancient sources place him even earlier. He is the designated author of the *Iliad* and the *Odyssey*.[92]

In the *Odyssey*, Homer gives a description of Ogygia, the island home of the nymph Calypso. She is also referred to as 'Atlantis' as she was the daughter of Atlas, the Titan.

Ogygia, like the island Atlantis, is situated in the *"far distant deep"* of the Atlantic Ocean (or simply Oceanus as Homer referred to it). Calypso lived in a cave on the island and *"Round her cave there was a thick wood of alder, poplar, and sweet-smelling cypress trees, wherein all kinds of great birds had built their nests—owls, hawks, and chattering sea-crows that occupy their business in the waters. A vine loaded with grapes was trained and grew luxuriantly about the mouth of the cave; there were also four running rills of water in channels cut pretty close together, and turned hither and thither so as to irrigate the beds of violets and luscious herbage over which they flowed. Even a god could not help being charmed with such a lovely spot,... ."*[93]

This, whilst not directly linked to Atlantis as described by Plato, does display a number of similarities to the description given by Plato and, furthermore, Atlantis/Calypso lived on the island.

3.3.2   Strabo

Strabo was a Greek historian and geographer who lived from about 64/63 BC to about AD 24.

In Book 1, Chapter 2, Section 18[94] of his *Geographica* Strabo, whilst criticising another author, Polybius, for disputing that Ogygia was in the Atlantic, recalls certain of Homer's statements with regard to Ogygia. Strabo says, albeit critically of Polybius, *"Such are the words of Polybius, and what he says is in the main correct. But when he demolishes the argument that places the wanderings of Odysseus on Oceanus, and when he reduces the nine days' voyage and the distances covered thereon to exact measurements, he reaches the height of inconsistency. For at one moment he quotes the words of the poet: 'Thence for nine whole days was I borne by baneful winds'; and at another moment he suppresses statements. For Homer says also: 'Now after the ship had left the river-stream of Oceanus'; and 'In the island of Ogygia, where is the navel of the sea,' going on to say that the daughter of Atlas lives there; and again, regarding the Phaeacians, 'Far apart we live in the wash of the waves, the farthermost of men, and no other mortals are conversant with us.' Now all these incidents are clearly indicated as being placed in fancy in the Atlantic Ocean ... ."*[95]

It is apparent from this that Strabo accepted the existence of the island of Ogygia, situate far away in the Atlantic Ocean.

3.3.3   Pliny the Elder

Pliny the Elder[96] mentions that, according to *"Timaeus the Historiographer"* within the ocean, *"at Six Days' sailing from Britannia, is the Island Mictis, in which White Lead* [tin] *is produced, and that the Britanni sail thither in Wicker Vessels, sewed round with Leather."*[97] He does not specifically mention

Ogygia but considering what other writers, in particular Plutarch, wrote there might be some link.

3.3.4   Plutarch

Plutarch, who lived from about 46 CE to 120 CE, was a Greek historian, biographer, and essayist. In his work *"Moralia"* he describes the island of Ogygia but goes beyond what Homer says, providing a more detailed description of where it was and the surrounds.

He states[98] *"An isle, Ogygia, lies far out at sea, a run of five days off from Britain as you sail westward; and three other islands equally distant from it and from one another lie out from it in the general direction of the summer sunset."*

The distance from Rockall to Britain is approximately 470 km (or 290 miles or about 250 nautical miles).[99] One does not know the conditions under which the boat (which took five days to reach Ogygia) was sailing but it is known that the prevailing winds in that area are westerly. A boat sailing from Britain to Ogygia/Hatton Rockall would, as such, have generally been sailing against the wind. McGrail,[100] states that a journey from Alexandria to Cyprus, which was also a distance of 250 nautical miles, was recorded as having taken 6.5 days in foul weather.

That being the case, it is possible that a journey by boat, from Britain to the Hatton Rockall area, could have taken about five days.

Plutarch also mentions, similarly to Plato, the *"great mainland, by which the great ocean is encircled, while not so far from the other islands, is about five thousand stades from Ogygia,... ."*

3.3.5   Olaus Rudbeck

Olaus Rudbeck, a Swedish scientist, writer and professor of medicine at Uppsala University, was born in 1630 and died in 1702. He is known, primarily, for his contributions in two fields, namely human anatomy and linguistics. He, however, also wrote a 3 000-page treatise in four volumes called *Atlantica* (also known as *Atland eller Manheim* in Swedish). He, also, proposed that Ogygia and Atlantis were the same place.[101].

From certain descriptions given in Homer's *Odyssey,*[102] Rudbeck attempted to calculate the latitudinal position of Ogygia. According to Zhirov,[103] Rudbeck estimated that it lay between approximately 51° N and 64⁰ N whilst, according to Sprague de Camp, Rudbeck calculated[104] that it lay *"between the latitudes of Mecklenburg, Germany and Vinililand, Sweden."* Mecklenburg is a historical region in northern Germany situate at about 53⁰ N,[105] and Stockholm, Sweden is at about 59⁰ N.

Rockall Islet lies at approximately 57⁰ N[106] while the entire Hatton Rockall area, or plateau, lies between, approximately, 53⁰ N and 59⁰ N[107].

3.3.6   Conclusion

I therefore propose that a sub-aerial Hatton Rockall could have been the island that has also been referred to by the name of Ogygia and, consequently, that Atlantis and Ogygia were, probably, one and the same place.

3.4   Other, unnamed, islands

3.4.1   *Pseudo-Aristotle*

de Mirabilibus Auscultationibus, or, translated into English, *On Marvellous Things Heard*,[108] is a work written by a Pseudo-Aristotle. (It is not known who actually wrote it but it was originally, erroneously, attributed to Aristotle.) It is uncertain when this was written but some scholars consider it to contain a core of early material from the Hellenistic period – the period just after Plato.[109] Although the name reflects a claim to be about things "heard," it has been suggested that it is more likely that the contents were from things read, and copied, from books.

What is important in this work is the statement[110] that *"In the sea outside the Pillars of Heracles they say that a desert* (also translated as 'desolate'[111]) *island was found by the Carthaginians, having woods of all kinds and navigable rivers, remarkable for all other kinds of fruits, and a few days⊠ voyage away ...."*

This description is remarkably similar to that given by Plato of Atlantis, even to the extent of describing the island as being outside the Pillars of Heracles. Whilst some might suggest that this description merely follows Plato, that is not so. No indication could have been given, if this statement was merely a copy of what Plato wrote, regarding the distance of this island from the Pillars. Plato, unlike the Pseudo-Aristotle, does not give an indication of the distance of the island.

3.4.2   *Proclus*

Proclus Lycaeus (February 412 – 17 April 485 CE), was a Greek Neoplatonist philosopher. The majority of his works are commentaries on dialogues of Plato,[112] one of which is a *Commentary on Plato's Timaeus*. In this Proclus, when commentating on Plato's statement that "… *this island was greater than both Libya and Asia together, and afforded an easy passage to other neighbouring islands* …" wrote[113]:

"*That such and so great an island once existed, is evident from what is said by certain historians respecting what pertains to the external sea. … according to them, there were seven islands in that sea, … and also three others of an immense extent … the magnitude of* [one of] *which was a thousand stadia. They also add, that the inhabitants of it preserved the remembrance from their ancestors, of the Atlantic island which existed there, and was truly prodigiously great; which for many periods had dominion over all the islands in the Atlantic sea, and was itself likewise sacred to Neptune*[Neptune was also known by the name of Poseidon]."

We do not know, unfortunately, to which historians Proclus was referring but, whoever they were, it is apparent that the main "Atlantic island" had disappeared and that there remained (after Atlantis's destruction) ten other islands.

This, given that islands only disappear if they sink or are inundated, gives credence to my proposition that Hatton Rockall gradually sank, leaving the high points as remaining islands - that is, at least, until they all ultimately sank and disappeared.

As an aside, if one looks at the region on NOAA's Bathymetric Data Viewer[114] (with the bathymetric contours option being selected), it appears that there could have been, in the region from Hatton Rockall to the Faeroes, at least around nine islands if the sea level was at, what is now, the 500 m contour level.

3.5   Hesperides

In Greek mythology the Hesperides are nymphs.[115] The deities of the evening and golden light of sunset,[116] who tend a verdant and blissful garden on an island in the far west[117] or north.[118] In the garden, the Hesperides are said to have guarded the golden apples of Hera.

There are various stories as to the paternity of the Hesperides but according to Diodorus Siculus their father was Atlas. In his *Library of History*, Book 4, 27.1 he wrote, "*In the country known as Hesperitis there were two brothers whose fame was known abroad, Hesperus and Atlas. These brothers possessed flocks of sheep which excelled in beauty and were in colour of a golden yellow, this being the reason why the poets, in speaking of these sheep as mela, called them golden mela. Now Hesperus begat a daughter named Hesperis, whom he gave in marriage to his brother and after whom the land was given the name Hesperitis; and Atlas begat by her seven daughters, who were named after their father Atlantides, and after their mother, Hesperides.[119]*"

In the early versions of the myth, the Hesperides were also an island (or islands) specifically set in the Ocean, the Atlantic.[120]

Commonly there are three Hesperides although, depending upon the source or writer, this number can vary between three and seven.[121] Calypso the daughter of Atlas, to whom I have already referred, is sometimes included within the Hesperides. The names of the Hesperides varied depending upon the writer and the number of Hesperides, however one name which is consistent across virtually all writers, is Erytheia. Erytheia was the mother of Eurytion, the herdsman of Geryon who, in the story of Heracles and his labours, lived on the island of Erytheia.[122]

So, in the Hesperides, we have a link to Atlas, his daughters known as Atlantides, who lived on an island in the far west with a verdant garden which, similarly to Avalon,[123] had apples.

Insofar as this issue of 'apples' is concerned Atlas, according to Plato, had a twin brother whose name in Greek was Eumelus. (His, alternate, non-Greek name was Gadeirus.). The name Eumelus may possibly be broken down into two aspects: 'Eu' and 'melus'. According to the *Shorter Oxford English Dictionary on historical principles*[124] the prefix 'eu' (generally in Greek derived words) conveys a sense of 'good' or 'well'. 'Melus' on the other hand may be translated as "apple".[125] If they are put together the name may, therefore, be translated as "Good (possibly beautiful?) Apple".[126]

The Hesperides, have also been linked with the Islands of the Blessed or the Fortunate Isles.

3.6   Hyperborean Island

Diodorus Siculus (who knew of the existence of the island of Britain and distinguishes Hyperborea from Britain) wrote in Chapter 47 of Book V of his *Library of History* (Professor Oldfather's translation in the Loeb Classical Library[127]) "*… we feel that it will not be foreign to our purpose to discuss the legendary accounts of the Hyperboreans. Of those who have written about the ancient myths, Hecataeus and certain others say that in the regions beyond the land of the Celts there lies in the ocean an island no smaller than Sicily. This island, the account continues, is situated in the north and is inhabited by the Hyperboreans, who are called by that name because their home is beyond the point whence the north wind (Boreas) blows; and the island is both fertile and productive of every crop, and since it has an unusually temperate climate it produces two harvests each year.*" and "*The Hyperboreans … are most friendly disposed towards the Greeks, and especially towards the Athenians and the Delians, who have inherited this good-will from most ancient times. The myth also relates that certain Greeks visited the Hyperboreans and left behind them there costly votive offerings bearing inscriptions in Greek letters. And in the same way Abaris, a Hyperborean, came to Greece in ancient times and renewed the good-will and kinship of his people to the Delians.*"

Diodorus relies (in the main) on a 'Hecataeus' for his description of Hyperborea. The Hecataeus to whom he refers is, probably, Hecataeus of Abdera,[128] a Greek historian who lived during the 4th century BCE. Hecataeus's major contributions were his ethnographical works, one of which was entitled '*On the Hyperboreans*'. This work was comprised of, at least, two books. Unfortunately both of these books have been lost but fragments of them have survived in references to them in other subsequent works – such as the *Library of History*.

Diodorus' reference to an island "*beyond the land of the Celts*" raises the question as to where the "*land of the Celts*" was. Gaul is often regarded as being the 'land of the Celts' however the Celts in fact spread as far as Ireland.

To know whether Diodorus could have been referring to Ireland we need to know when the Celts arrived in Ireland.

There are various theories as to when the Celts settled in Ireland. These theories place the Celts arrival there between, as early as, the 7th century BCE and, as late as, the 1st century BCE. This period covers the times of both Hecataeus of Abdera and Diodorus Siculus.[129] As such, the reference by Diodorus Siculus to *"the land of the Celts"* could include a reference to Ireland and Britain, which would also have been a Celtic land at the time. I am corroborated in this conclusion that the *"lands of the Celts"* included Britain by the work *Circuit of the Earth* which is attributed to Pseudo Scymnus (or Pausanias of Damascus as he has also been called). This was probably written in about 100 BCE.[130] In it Pseudo Scymnus refers to *"riverborne tin from* <u>Keltike</u>*"* (alluvial tin was found in the early days in Devon and Cornwall, England) and also states that *"… the country called Keltike … [is] … the greatest nation in the West."* and further that *"… the* <u>Kelts</u> *have from the west wind up to the summer sunset …"*.[131]

Diodorus described the Hyperboreans' island as being *"not smaller than Sicily"* – implying that it could be larger. Sicily, although smaller than Ireland in surface area, is the largest island in the Mediterranean and would logically, for people living around the Mediterranean, be used to convey an island of large size. It is therefore reasonable to assume that the Hyperboreans' island was considered to be a large island.

Therefore, once again, there is a reference to a large island found to the west, in the Atlantic Ocean, which had a temperate climate, was very fertile and produced two harvests a year.

Also of importance for my hypothesis is the fact that Hyperborea was described as being situate in the far north of the Atlantic Ocean.

3.7   Hy-Brasil

Hy-Brasil, or Brasil as it is also known,[132] is generally accepted as being a purely mythical, or phantom, island. It first appeared on a 'map'[133] as early as 1325 when it was marked as "Bracile" and continued to appear on various other maps until the second half of the 1800's. On some maps it has appeared twice[134] and has, to a certain extent, moved around, but the one position to which it has remained faithful is in the Atlantic ocean to the west of Ireland.[135,136]

There have been various explanations as to how the name arose one of which is that it derives from Old Irish: Í: island; *bres*: beauty, worth, great, mighty.[137] This latter description would be consistent with the "Blest Isle". According to the author William H Babcock[138] the name is far older than when it first appeared on a map in 1325 and has been associated with the '*Fortunate Islands.*'

In virtually all portrayals of the island it lacks any significant topographical details, contrary to what would normally be expected, and is often portrayed as round. The one other characteristic which is sometimes displayed is a 'channel' through the middle of the island, effectively dividing it into two.

It appears, from these, that Hy-Brasil might have been an actual island, remembered only vaguely,

> ➢ off the west coast of Ireland

> ➢ which had a channel running through it and

> ➢ which warranted a 'commendatory' appellation.

If Hy-Brasil was a residual subaerial Hatton Rockall it is possible that the central plain had already submerged leaving the two highlands, Hatton High and Rockall High, as two separate, but closely spaced, islands.

On its own, Hy-Brasil must remain a phantom, mythical island but, if there is anything to my hypothesis, it, might have a spark of life breathed into it again and provide some corroboration for my proposition.

3.8   Avalon

Another mythical island situate to the west and which comes from Celtic myth is Avalon. The name Avalon probably comes from the Welsh *Ynys Afallon. Afallon* is, in turn, probably derived from *afal,* meaning apple.[139]

As a legendary island featuring in Arthurian legend it first appears in Geoffrey of Monmouth's 1136 CE pseudo-historical account *Historia Regum Britanniae*. In his later work *Vita Merlini,* or The Life of Merlin, he called it *Insula Pomorum,* the "Isle of Apples".[140,141] There he gave the following description of the island "*The island of apples which men call "The Fortunate Isle" gets its name from the fact that it produces all things of itself; the fields there have no need of the ploughs of the farmers and all cultivation is lacking except what nature provides. Of its own accord it produces grain and grapes, and apple trees grow in its woods from the close-clipped grass. The ground of its own accord produces everything instead of merely grass, and people live there a hundred years or more. There nine sisters rule by a pleasing set of laws those who come to them from our country.*"[142,143] (Could the "*nine sisters*" be a link to the Hesperides?)

This description appears, again, to be consistent with the 'glowing' terms used by Plato in his description of Atlantis and also with the description of Ogygia.

3.9   Conclusion

From the above it appears that there were ancient memories of an island, or islands, which existed out in the extreme west, in the Atlantic Ocean.

If this is accepted (and, after reading the next section it is accepted that Hatton Rockall meets the description given by Plato of Atlantis) then it appears plausible that Hatton Rockall may have sunk gradually (albeit very fast by generally accepted geological time frames) and over millennia to its present depth.

I shall deal with a proposed mechanism for the submersion in section 13.

SECTION FOUR:
PLATO AND HIS DESCRIPTION OF ATLANTIS

4.1   Introduction

In this section I shall, firstly, provide a very brief introduction to Plato. I shall then consider the descriptions of Atlantis given by Plato, collate the information and extract the salient points of his description. It is against these salient points that the present evidence must be weighed.

Plato's two dialogues the *Timaeus* and the *Critias*,[144,145] are accepted as the original descriptions of Atlantis and I shall quote liberally from them.

In recording the salient points of Plato's description I use a 'point form' recitation of the descriptions. This is not, I recognise, conducive to a smooth flow narrative but I do it for the sake of clarity and brevity.

4.2   Plato

Plato lived from about 428 to 348/347 BCE, was a philosopher in classical Greece, a writer of philosophical dialogues and a mathematician.[146] He is generally considered to be the greatest philosopher in the history of western civilisation.

According to Taylor[147] Plato came from a most illustrious and distinguished pedigree. On his father's side his ancestors were said to go back to the ancient kings of Athens and even beyond to the god Poseidon.[148] His ancestors on his mother's side are also illustrious but somewhat more historically ascertainable. One of his illustrious ancestors was Solon, an Athenian lawmaker and statesman.[149]

Plato wrote in a period when Greek culture was moving from an oral based culture to a literate culture.[150] According to Gibson[151] Plato wrote in a traditional oral style and, while doing so, argued for an alternate, or competing, branch of the oral tradition which still preserved and transmitted information.[152] In his written works Plato very rarely mentioned himself[153] and his works were often presented in the form of dialogues, thereby following the older and more traditional oral format.

In addition to his writings Plato gave lectures but none of these lectures have been recorded in writing.[154] Plato, it would appear, generally had a great reticence about recording serious, philosophical, matters in writing.[155] In his *Seventh Letter*[156] he wrote " ... *every serious man in dealing with really serious subjects carefully avoids writing, ...*".[157] The reason for this, also set out in his *Seventh Letter,* was in order not " ... *to expose them to unseemly and degrading treatment*".[158]

For Plato, the truth was extremely important and, for him, philosophy meant an active personal pursuit of truth and goodness.[159]

Notwithstanding his stated reticence to putting a 'serious subject' in writing he had also written earlier in his life, in his *Phaedrus* (through the voice of Socrates), that "... *The gardens of letters he will, it seems, plant for amusement, and will write, when he writes, to treasure up reminders for himself, when he comes to the forgetfulness of old age, and for others who follow the same path, and he will be pleased when he sees them putting forth tender leaves.* ".[160] His reticence to put things in writing was, it is apparent, not absolute. He was, sometimes, willing to set things down in writing for others following.

As regards the story of Atlantis itself, this came down to Plato through various ancestors, the earliest of whom, Dropides, had been a friend of Solon.[161] It was Solon who originally learned of Atlantis from the Egyptians. Plato, through the mouth of Critias, explains: "*Solon, who was intending to use the tale for his poem, enquired into the meaning of the names, and found that the early Egyptians in writing them down had translated them into their own language, and he recovered the meaning of the several names and when copying them out again translated them into our language. My great-grandfather, Dropides, had the original writing, which is still in my possession, and was carefully studied by me when I was a child.*"[162, 163]

The *Timaeus*, in which the first mention of Atlantis is made, was written after Plato's work the *Republic*[164] and can be dated to somewhere between 360 BCE and Plato's death in 347 BCE.[165] Plato died at about age 80. This means that he could not have been younger than 67 when he wrote the *Timaeus* and the *Critias* (which followed the *Timaeus*). It follows that, notwithstanding Plato's general aversion to setting out serious subjects in writing, he could have done it *"to treasure up reminders for himself, when he comes to the forgetfulness of old age"* or *"for others who follow the same path …"*.

It appears to me (whilst acknowledging that I have not studied Plato) to be anomalous that Plato, one of the greatest western philosophers, would have included the story of Atlantis in the *Timaeus* (holding it out as fact) simply for the sake of creating a fictional, idyllic, scene for <u>part</u> of the discussions. The *Timaeus*, is a dialogue (in essence, a debate) about the creation of the cosmos by the 'demiurge'. The story of Atlantis forms a very minor part of the dialogue and it adds nothing significant to it.

Given Plato's philosophy, his active pursuit of the truth and his links to Solon, I propose that his description of Atlantis should not be rejected out of hand simply because it conflicts with generally accepted history. I also hope that you will, at the end, be persuaded that it is implausible that Plato's description of the physical Atlantis was a figment of his imagination.

4.3   Atlantis

4.3.1   *A description of Atlantis – its size, where it was and when and how it disappeared.*

The following excerpts (all of which are from the Jowett translation) describe Atlantis.

In the *Critias* it is stated that:

- ➢ the Atlanteans were *"those who dwelt outside the Pillars of Herakles"*[166]

- ➢ Atlantis *"was an island greater in extent than Libya and Asia …* "[167]

- ➢ "[Atlantis] *sunk by an earthquake,* [and] *became an impassable barrier of mud to voyagers sailing from hence to any part of the Ocean."*[168] and

> *"... nine thousand was the sum of years which had elapsed since the war which was said to have taken place ..."* and

> *"Many great deluges have taken place during the nine thousand years, for that is the number of years which have elapsed since the time of which I am speaking ... ."*

The period of 9 000 years is also, obliquely, mentioned in the *Timaeus*. In the *Timeaus* it is stated that:

> *"This power* [the Atlanteans] *came forth out of the Atlantic Ocean, for in those days the Atlantic was navigable; and there was an island situated in front of the straits which are by you called the Pillars of Heracles;*[169] *the island was larger than Libya and Asia put together,*

> [Atlantis] *... was the way to other islands, and from these you might pass to the whole of the opposite continent which surrounded the true ocean; for this sea which is within the Straits of Heracles is only a harbour, having a narrow entrance, but that other is a real sea, and the surrounding land may be most truly called a boundless continent."*

> *"But afterwards there occurred violent earthquakes and floods, and in a single day and night of misfortune all your warlike men in a body sank into the earth, and the island of Atlantis in like manner disappeared in the depths of the sea. For which reason the sea in those parts is impassable and impenetrable, because there is a shoal of mud in the way; and this was caused by the subsidence of the island."*[170]

Some writers have suggested that this nine thousand year period should, in fact, be regarded as a reference to nine hundred years and that Plato inflated his numbers by a factor of ten. I admit that some of the figures mentioned by Plato are 'hard to swallow' but this is not, in my view, a ground to simply assume that Plato made a mistake nor deliberately inflated his figures in these two works. Plato would have been aware that recorded Egyptian history went back more than 900 years prior to him; after all Solon had lived about 200 years prior to him.

Insofar as it might be suggested that Plato had merely wanted to convey that the destruction of Atlantis had occurred many, many thousands of year ago. Had this been his intent he would have used the Greek term 'myriad'. 'Myriad' was the largest specific number in ancient Greek, meaning 10 000, but could also, depending upon the context in which it was used, mean innumerable or infinite.

That Plato was aware of the term, or concept of, a myriad is apparent from the fact he used it in *The Laws*. There he described a deluge (in which the world of men was destroyed leaving only a small portion of the population surviving) as having occurred μυριάκις μύρια / myriákis mýria. This has been translated as *"thousands upon thousands of years"*,[171] *"untold tens of thousands of years"*[172] and *"ten thousand times ten thousand years"*. It is apparent, from the context and manner in which it was used there, that he intended to convey that that particular deluge had occurred an extremely long time ago - it was the devastation wrought and not the exact date of the deluge which was important.

Plato did not, in either the Critias or in the Timaeus, use the term 'myriad' but used the specific figure of nine thousand years. It is therefore apparent that Plato, when referring to the period of nine thousand years, was not simply referring to a time untold ages ago but rather to a specific time period. It is to events in that time period that we must look.

That said, it is improbable, as I have set out in section 11.2.1, that the story was told on the exact anniversary of the submersion of Atlantis.

4.3.1.1   Analysis and extraction of details

Summarising Plato's statements about Atlantis:

> ➢ Atlantis was a large island in the Atlantic ocean (I do not regard the statement that it was " *greater in extent than Libya and Asia*" as reliable, as one does not know what was envisaged by Plato as encompassing Libya and Asia, but will deal with this aspect in greater detail in section 5.4.)

> ➢ which around, approximately, 9400 BCE,[173]

➢ was rocked by violent earthquakes and floods,

➢ subsided into, or was covered by, the sea and

➢ the sea was, as a result of the catastrophe, rendered impassable.

If there is any doubt as to whether the Greeks at the time of Plato knew of the Atlantic Ocean Herodotus (who preceded Plato in time and who is known as the "Father of History") stated in his work *The Histories*[174] that "*The Caspian is a sea by itself, having no connection with any other. The sea frequented by the Greeks, that beyond the Pillars of Hercules, which is called the Atlantic, and also the Erythraean,*[175] *are all one and the same sea.*"[176] (I discuss and expand upon this in Sections 5.1 and 5.2).

Also, as an aside, it is interesting to note that before the 'New World' was, ostensibly, known to Europe, Plato mentions being able to cross the ocean, by way of "*other islands*" to "*the opposite continent*[177] *which surrounded the true ocean.*" and recognised that the Mediterranean sea is a mere harbour compared with the ocean outside the Pillars of Heracles.[178] (These issues will also be considered further in sections 5 and 11.)

4.3.2   *Physical features, climate and environment of Atlantis*

As to the physical features, climate and environment of the island itself Critias said that

➢ "*Looking towards the sea, but in the centre of the whole island, there was a plain which is said to have been the fairest of all plains and very fertile. Near the plain again, and also in the centre of the island at a distance of about fifty stadia*[179], *there was a mountain not very high on any side.*"[180]

➢ on this central mountain were "*two springs of water from beneath the earth, one of warm water and the other of cold …*"[181]

➢ "*… the island itself provided most of what was required by them for the uses of life*"[182]

> *"There was an abundance of wood for carpenter's work, and sufficient maintenance for tame and wild animals. Moreover, there were a great number of elephants in the island; for as there was provision for all other sorts of animals, both for those which live in lakes and marshes and rivers, and also for those which live in mountains and on plains, so there was for the animal which is the largest and most voracious of all. Also whatever fragrant things there now are in the earth, whether roots, or herbage, or woods, or essences which distil from fruit and flower, grew and thrived in that land; also the fruit which admits of cultivation, both the dry sort, which is given us for nourishment and any other which we use for food—we call them all by the common name of pulse, and the fruits having a hard rind, affording drinks and meats and ointments, and good store of chestnuts and the like, which furnish pleasure and amusement, and are fruits which spoil with keeping, and the pleasant kinds of dessert, with which we console ourselves after dinner . "[183]*

> *"… they had fountains, one of cold and another of hot water, in gracious plenty flowing; and they were wonderfully adapted for use by reason of the pleasantness and excellence of their waters."[184]*

> *"The whole country was said by him to be very lofty and precipitous on the side of the sea, but the country immediately about and surrounding the city was a level plain, itself surrounded by mountains which descended towards the sea; it was smooth and even, and of an oblong shape, extending in one direction three thousand stadia,[185] but across the centre inland it was two thousand stadia.[186] This part of the island looked towards the south, and was sheltered from the north. The surrounding mountains were celebrated for their number and size and beauty … "[187]*

As a point to note, the sentence "*This part of the island looked towards the south, and was sheltered from the north.*" is translated in the Perseus Tufts version as "*And this region, all along the island, faced towards the South and was sheltered from the Northern blasts.*"[188] I understand the reference to

"Northern blasts" to be a reference to strong and, presumably, unpleasant winds from the north. This is an important elaboration which I shall discuss in greater detail in section 6.2.

> The plain "... *was for the most part rectangular and oblong ... It received the streams which came down from the mountains ...*"[189]

> "*Twice in the year they gathered the fruits of the earth — in winter having the benefit of the rains of heaven, and in summer the water which the land supplied by introducing streams from the canals.* "[190]

Mining was also conducted on the island. It was said that

> "*... they dug out of the earth whatever was to be found there, solid as well as fusile, and that which is now only a name and was then something more than a name, orichalcum, was dug out of the earth in many parts of the island, being more precious in those days than anything except gold.* "[191]

Orichalcum was a metal, described by Critias as, flame coloured[192] or flashing with a red light.[193]

4.3.2.1   Analysis and extraction of details

From the above it is apparent that Atlantis

> was a large island situated in the Atlantic Ocean,

> had a central plain

>> which looked to the south and

>> was surrounded by mountains

>>> which were lofty and

>>> which descended precipitously to the sea on the side away from the plain and

> certain of which sheltered the plain from the northerly winds

> had, in the centre of the island, a hill or small mountain

> had lush natural (and cultivated) growth and vegetation and, presumably, a temperate or warm climate

> received most of its rain in winter

> had hot springs and

> had minerals to be mined, one of which (orichalcum) presumably was scarce as, by the time of the telling of the story, it was merely a name.

4.3.3   *The plain of Atlantis*

Further details with regard to the plain itself were also given. I quote below from both the Jowett and Perseus Tufts translations, Jowett first and then the Perseus Tufts.

"Now as a result of natural forces, together with the labours of many kings which extended over many ages, the condition of the plain was this. It was originally a quadrangle, rectilinear for the most part, and elongated; and what it lacked of this shape they made right by means of a trench dug round about it. Now, as regards the depth of this trench and its breadth and length, it seems incredible that it should be so large as the account states, considering that it was made by hand, and in addition to all the other operations, but none the less we must report what we heard: it was dug out to the depth of a plethrum and to a uniform breadth of a stade, and since it was dug round the whole plain its consequent length was 10,000 stades.[194] It received the streams which came down from the mountains and after circling round the plain, and coming towards the city on this side and on that, it discharged them thereabouts into the sea."[195]

"I will now describe the plain, as it was fashioned by nature and by the labours of many generations of kings through long ages. It was for the most part rectangular and oblong, and where falling out of the straight line followed the circular ditch. The depth, and width, and length of this ditch were incredible, and gave the impression that a work of such extent, in addition to so many others, could never have been artificial. Nevertheless I must say what I was told. It was excavated to the depth of a hundred feet, and its breadth was a stadium everywhere; it was carried round the whole of the plain, and was ten thousand stadia in length."[196]

"Further inland, likewise, straight canals of a hundred feet in width were cut from it through the plain, and again let off into the ditch leading to the sea: these canals were at intervals of a hundred stadia, and by them they brought down the wood from the mountains to the city, and conveyed the fruits of the earth in ships, cutting transverse passages from one canal into another, and to the city."[197]

"And on the inland side of the city channels were cut in straight lines, of about 100 feet in width, across the plain, and these discharged themselves into the trench on the seaward side, the distance between each being 100 stades."[198]

"Near the plain again, and also in the centre of the island at a distance of about fifty stadia, there was a mountain not very high on any side. … and breaking the ground, [the god Poseidon] enclosed the hill in which she dwelt all round, making alternate zones of sea[199] *and land larger and smaller, encircling one another; there were two of land and three of water, which he turned as with a lathe, … "*[200]

"Bordering on the sea and extending through the centre of the whole island there was a plain, which is said to have been the fairest of all plains and highly fertile; and, moreover, near the plain, over against its centre, at a distance of about 50 stades, there stood a mountain that was low on all sides. … and to make the hill whereon she dwelt impregnable he broke it off all round about; and he made circular belts

of sea and land enclosing one another alternately, some greater, some smaller, two being of land and three of sea, which he carved as it were out of the midst of the island;…" [201].

4.3.3.1   Analysis and extraction of details

From this it is apparent that

> ➤ there was a very deep trench, or channel, which went around the edge of the plain

> ➤ the channel looked too large to have been man-made but the storyteller had been informed that it was man-made,

> ➤ there were other canals which crossed the plain and

> ➤ there was a hill in the centre of the plain with moats around it.

4.3.4   Atlantis and Gades

Finally, it is also stated, in the *Critias*, that Poseidon divided the island into ten portions and allocated a portion to each one of his ten male children. "… *he gave to the first-born of the eldest pair* ("Atlas") *his mother's dwelling and the surrounding allotment … [and] … To his twin brother … the extremity of the island towards the Pillars of Heracles, facing the country which is now called the region of Gades in that part of the world …."*

The Perseus Tufts, or W.R.M. Lamb, translation reads: "… *he assigned to the first-born of the eldest sons his mother's dwelling and the allotment surrounding it … And … his younger twin-brother, who had for his portion the extremity of the island near the Pillars of Heracles up to the part of the country now called Gadeira …."* [202]

As can be seen the place name *"Gades"* (as it appears in the Jowett translation) is translated in the Perseus Tufts translation as *"Gadeira"*. A detailed consideration of what, or where, Gadeira was is given in section 9.

4.3.4.1   Analysis and extraction of details

From this we may conclude that, at the extremity of the island, which faces towards the Pillars of Heracles,

> ➢ there was another land, or region, which was then known by the name of Gades / Gadeira.

Gades is widely supposed to be the modern day Cadiz in Spain however I shall put forward an alternate proposition when I deal with this in Section 9.

4.4   Conclusion

In summary therefore Atlantis was a large island in the Atlantic Ocean with a temperate climate which received most of its rain in winter. It had a central plain which was largely rectangular and aligned, more or less, North/South. The plain was divided up by a number of channels criss-crossing it, had a small mountain in the middle of it and had a deep channel running around its edge. Around the plain were tall mountains which were particularly steep on the seaward side and a portion of the island faced a region then known as Gades/Gadeira.

These salient points must be compared to the know geography of the Hatton Rockall area but, before doing so, I shall consider what Plato contemplated when he used the names, or descriptions, which have been translated as the 'Pillars of Heracles', the 'Atlantic', 'continent' and that Atlantis was larger, or greater, than Libya and Asia combined.

The issues of the Pillars of Heracles and the Atlantic are particularly important as, without having a degree of certainty about these, one cannot place Atlantis.

SECTION FIVE:
WHAT DID PLATO ENVISAGE BY THE TERMS 'THE PILLARS OF HERACLES', 'THE ATLANTIC' AND 'CONTINENT' AND THAT ATLANTIS WAS LARGER THAN LIBYA AND ASIA

5.1   The Pillars of Heracles

The Pillars of Heracles are critical in the construction of any theory relating to the location of Atlantis. This is so because, according to Plato, Athens (and Egypt) were situate within the Pillars whilst Atlantis was situate outside of them. (Atlantis lay *"in front of the mouth*"[203] or *"beyond the Pillars of Heracles* "[204].) Any proposed location of Atlantis must therefore be located (relative to Athens) beyond the Pillars.

The question to be considered and answered in this section is therefore: 'Where did <u>Plato</u> consider that the Pillars of Heracles were situated?' (A different and, at this stage, erroneous question to ask is: Where, in fact, were the Pillars situated?)

Given that Plato was a philosopher and that these stories were aimed at his audience, it can be safely assumed that they were aimed (at least primarily) at relatively educated adults. The stories are, however, also capable of being understood on a simple, superficial, level as well. It is probable, therefore, that Plato would have made use of current terms and knowledge in the story and would not have used obscure, or abstruse, terms.

In order to answer the above question we should first consider what Plato said about the Pillars, the actual geography of the Mediterranean Sea, relevant writings of writers earlier than, or concurrent with, Plato (to determine the, then, currently accepted meanings of the term), the writings of writers after Plato (but only insofar as they may throw some light on what Plato wrote) and, finally, assess what is the most likely, and probable, position of the Pillars as referred to by Plato.

5.1.1   *Plato's references to the Pillars*

Plato mentions the Pillars of Heracles both in the *Timaeus* and the *Critias* (*Timaeus*, 24e; *Critias*, 108e; 114b), and also in the *Phaeton* (109a / 109b). I shall use the Perseus Tufts translations. They read, respectively,

> *Timaeus* 24e "... *For it is related in our records how once upon a time your State stayed the course of a mighty host, which, starting from a distant point in the Atlantic ocean, was insolently advancing to attack the whole of Europe, and Asia to boot. For the ocean there was at that time navigable; for in front of the mouth which you Greeks call, as you say, 'the Pillars of Heracles,' there lay an island which was larger than Libya and Asia together; and it was possible for the travellers of that time to cross from it to the other islands, and from the islands to the whole of the continent ...".*[205].

Section 24e of *Timaeus* must, for a proper understanding of it, be read together with sections 25a and 25b, both of which give further insight into where Plato considered the Pillars to be. The relevant portions of these sections are quoted below – after the quote from the *Phaeton*.

> *Critias* 108e "... *since the war occurred, as is recorded, between the dwellers beyond the Pillars of Heracles and all that dwelt within them; which war we have now to relate in detail. It was stated that this city of ours was in command of the one side and fought through the whole of the war, and in command of the other side were the kings of the island of Atlantis, which we said was an island larger than Libya and Asia once upon a time, but now lies sunk by earthquakes and has created a barrier of impassable mud."*[206]

> *Critias* 114b " ... *who had for his portion the extremity of the island near the Pillars of Heracles up to the part of the country now called Gadeira ...* "[207]

> *Phaeton* 109a "*Secondly,*" said he, "*I believe that the earth is very large and that we who dwell between the Pillars of Hercules* [109b]

and the river Phasis[208] *live in a small part of it about the sea, like ants or frogs about a pond, and that many other people live in many other such regions."*[209]

The relevant portions of Sections 24a and 25b of the *Timaeus* read,[210] respectively, *"For all that we have here, lying within the mouth* (στόματος / stómatos) *of which we speak, is evidently a haven having a narrow entrance; but that yonder is a real ocean ..."*, and " *... of the* <u>lands here within the Straits</u> *they ruled over Libya as far as Egypt, and over Europe as far as Tuscany. So this host, being all gathered together, made an attempt one time to enslave by one single onslaught both your country and ours and the whole of the territory within the Straits* (στόματος / stómatos) *... ."*

There are two significant aspects to this:

➤ the use of the word στόματος / stómatos (which is translated as *"the Straits"* but the literal meaning of it is 'mouth') in conjunction with the Pillars of Heracles indicates that the Pillars were not mere pillars but rather that they are at, or formed, a mouth, or strait, in the sea and

➤ the fact that both Libya (which, it must be remembered, referred to, at least, northern Africa) as far as Egypt[211] and Europe as far as Tuscany (or Tyrrhenia) are referred as being within (i.e. on the same side of) the Straits.[212]

It is important to emphasise that, however one defines Libya, it must extend as far as Egypt – based on the statement that the Atlanteans ruled over *"Libya as far as Egypt ... "*

Those being accepted one may consider (accepting that Libya referred to northern Africa) within what mouth, or strait, did <u>both</u> northern Africa <u>as far as Egypt</u> and Europe as far as Tuscany lie?

Below is a map, Figure 5.1, showing the Mediterranean Sea, northern Africa and southern Europe. There are, essentially, only three straits which may be relevant – the Strait of Gibraltar, the Strait of Messina and the Strait of Sicily.

The Straits of Sicily and of Messina are, respectively, the open sea between Sicily and Tunisia (in North Africa) and the narrow strait between the eastern tip of the island of Sicily and the western tip of Calabria in southern Italy. Tuscany (which is on the north west of the 'boot' of Italy) is on one side of both of these straits whilst <u>Libya as far as Egypt</u> extends to the east beyond both Straits.

On this basis one may eliminate the Strait of Sicily (which is also hardly a 'mouth') and the Strait of Messina as being the 'mouth' to which Plato referred.

This leaves only the Strait of Gibraltar within which both Europe as far as Tuscany <u>and</u> Libya as far as Egypt are situate.

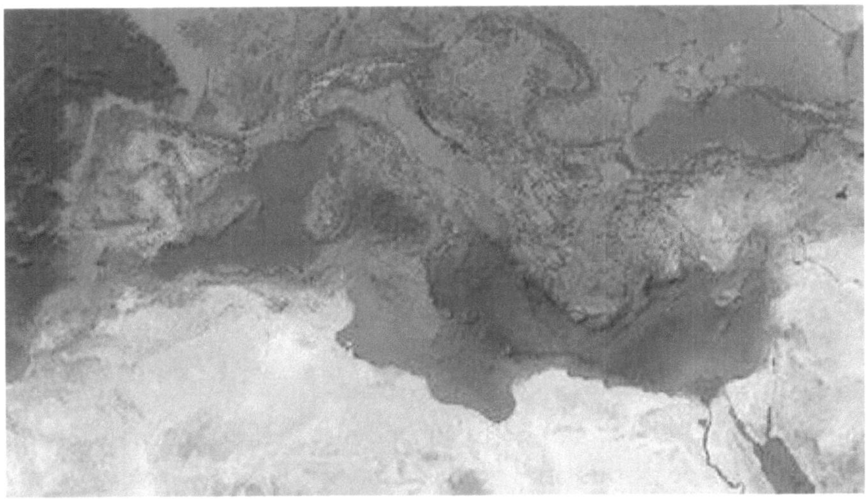

Figure 5.1 1 Map of the Mediterranean Sea and surrounding land. [213]

A final significant point to be reiterated and emphasised is that it is apparent from the description that Athens was within the Pillars of Heracles and the Atlanteans' island was outside the Pillars. That being so, Atlantis must have been in the Atlantic - as we now know it.

5.1.2   *Writers before Plato mentioning the Pillars*

5.1.2.1   Homer

Homer (late 8th or early 7th century BCE) is the attributed composer of the works the Iliad and the Odyssey. He does not appear to specifically mention the Pillars of Heracles but he does make it clear that the 'Okeanos' is beyond the limits of the known world and therefore, implicitly, beyond the Pillars of Heracles. Homer's Iliad states "... *I am faring to visit the limits of the all-nurturing earth, and Oceanus, from whom the gods are sprung ...*[214]" and "... *the great might of deep-flowing Ocean, from whom all rivers flow and every sea, and all the springs and deep wells ... .*"[215]

Other early writers, including Hesiod, Aeschylus and Aristophanes,[216] also mention the Okeanos and place it on the outer limits of the known world.

5.1.2.2   Hecataeus of Miletus

Hecataeus of Miletus (*circa*. 550 BCE – c. 476 BCE) was an early Greek historian and geographer. All his works have been lost but certain quoted fragments remain. He provides the first mention of the Pillars of Heracles.

His work *Periodos ges* (a comprehensive work on geography) described a journey beginning at the Straits of Gibraltar and, going clockwise (following the coast of the Mediterranean and Black Seas), and ending on the opposite, Moroccan, side.[217] The Pillars are referred to in Fragments 39, 41 and 356.[218]

5.1.2.3   Pindar

Pindar, (*circa*. 522 – c. 443 BCE) an Ancient Greek lyric poet places the Pillars (in his ode Isthmian 4, *For Melissus of Thebes Pancratium*) at the furthest limits stating "*Through their manly deeds they reached from home to touch the farthest limit, the Pillars of Heracles — do not pursue excellence any farther than that!*"[219]

In his *Nemean Ode* Pindar elaborates, indicating that the sea beyond the Pillars is trackless and, by implication, vast. He wrote "*.. it is not easy to cross the trackless sea beyond the Pillars of Heracles, which that hero and god set up as famous witnesses to the furthest limits of seafaring. He subdued the monstrous beasts in the sea, and tracked to the very end the streams of the shallows, where he reached the goal that sent him back home again, and he made the land known.*"[220]

He also wrote, in his Olympian Odes, of the '*Island of the Blessed*' in the Ocean[221].

Pindar, according to Strabo[222], also referred to the Pillars of Heracles as the "*the Gates of Gades*". Gades has generally[223] been accepted as referring to, what is now known as, Cadiz in southern Spain.

5.1.2.4 Hanno the Navigator

The periplus[224] attributed to Hanno the Navigator (a Carthaginian, or Phoenician, explorer of the sixth or fifth century BCE) supplies a description of a journey, which is generally accepted to have been down the west coast of Africa. It specifically mentions the "Pillars of Hercules" and sailing outside them.

5.1.2.5 Herodotus

Herodotus (*circa* 484– 425 BCE) specifically mentions the Pillars of Heracles and mentions them in the context of Okeanos. He wrote that Geryones "*... settled in the island called by the Greeks Erythea, on the shore of Ocean near Gadira, outside the Pillars of Heracles.*[225]" Here it is explicit that the Okeanos lies outside the Pillars.

Herodotus also, explicitly states, with regard to the seas, that "*... the Caspian sea. This is a sea by itself, not joined to the other sea. For that on which the Greeks sail*[226], *and the sea beyond the Pillars of Heracles, which they call Atlantic, and the Red Sea, are all one: but the Caspian is separate and by itself.*"[227] In this statement he considers three 'seas', the Mediterranean (the sea "on which the Greeks sail"), the Atlantic and the Erythraean. He specifically recognises that all these seas are (unlike the Caspian) linked.

It is, therefore, explicit that for Herodotus the Atlantic lies beyond the Pillars of Heracles and that both the Atlantic and the Erythraean Sea are linked and parts of a greater whole.

Herodotus also mentions[228] that a certain Colaeus[229] (who lived about 640 BCE) had travelled (or been driven by bad weather) beyond the Pillars of Heracles and had travelled to Tartessus,[230] a city and area generally accepted as being on the Atlantic coast of Spain.

5.1.2.6 Euripides

Euripides, who lived at much the same time as Herodotus (*circa.* 480 – 406 BCE), was a tragedian of classical Athens. Being an author of plays it follows, by necessary implication, that the terms and concepts he used must have been understandable to his audience.

He wrote, amongst many others, the play Hippolytus. In this play he referred to the Pillars of Heracles or, as he called them, the Pillars of Atlas. In the opening lines he sets the physical scene as being between the Black Sea and the Pillars. There he has the goddess Aphrodite say *"Mighty and of high renown, among mortals and in heaven alike, I am called the goddess Aphrodite. Of all those who dwell between the Euxine Sea[231] and the Pillars of Atlas and look on the light of the sun, ..."* [232,233] He clarifies that the Pillars of Atlas, as he calls them, were in the west by writing *"To the strand of the Daughters of the Sunset, The Apple-tree, the singing and the gold; Where the mariner must stay him from his onset, And the red wave is tranquil as of old; Yea, beyond that Pillar of the End That Atlas guardeth, would I wend; ..."*[234,235]

5.1.2.7   Pseudo-Scylax

Around the mid-4th century BCE another periplus, attributed to the so-called Pseudo-Scylax,[236] was written. It describes a voyage around the Mediterranean Sea and commences with *"I begin from the Pillars of Herakles in Europe until the Pillars of Herakles in Libya and until the tall Ethiopians[237]. The Pillars of Herakles are opposite each other and a day's sail apart. Beyond the Pillars of Herakles in Europe are many trading posts of the Carthaginians and clay/mud and flood tides and shoals."*[238]

It is interesting to note (in relation to the comments about Aristotle and the shallow and muddy Atlantic referred to below) that Pseudo-Scylax specifically mentions the mud and shoals in conjunction with the Phoenicians/Carthaginians.

5.1.2.8   Conclusion

It is apparent from these extracts that, prior to and at the time of Plato, the term the "Pillars of Heracles' was understood to refer to what is now known as the Strait of Gibraltar.

5.1.3   Relevant writers after Plato mentioning the Pillars

There is, in particular, one later writer who may throw light upon what Plato was describing when he referred to the Pillars of Heracles. That writer is Aristotle (384–322 BCE), he was a student of Plato's and, for that reason, may be able to throw some light on Plato's description.

5.1.3.1 Aristotle

Aristotle mentions the Pillars of Heracles a few times in his extant works. In his *Rhetoric* he writes, when discussing envy and, more particularly, what people do not envy, " *… for no man tries to rival those who lived ten thousand years ago, … nor those who live near the Pillars of Hercules; … .*"[239] He is clearly using the reference to the 'Pillars' as denoting a place far away and at the limits of the Greek world. This indication of an extreme western boundary is also visible in the following quotation from his work *Meteorology.* "*If we compute these voyages and journeys the distance from the Pillars of Heracles to India exceeds that from Aethiopia to Maeotis and the northernmost Scythians by a ratio of more than 5 to 3, as far as such matters admit of accurate statement. … But it is the sea which divides as it seems the parts beyond India from those beyond the Pillars of Heracles and prevents the earth from being inhabited all round.*"[240] Clearly he regards India and the Pillars as being on opposite sides of the, known, world.

Although it is apparent from this that Aristotle regarded the Pillars of Heracles as being at the western extreme of the known world there is another statement made by him which seems to jar with this. It is a statement often latched onto by others for whom the position at the Straits of Gibraltar conflicts with their theory.

Aristotle states, also in his work *Meteorology,* that "*Outside the Pillars of Heracles the sea is shallow owing to the mud, but calm, for it lies in a hollow.*"

This appears to be incongruous with the position of the Pillars at the Straits of Gibraltar and is not a description of the Atlantic as we know it, which is neither shallow nor calm. (Nor is it a description of the western Mediterranean, which is also not shallow either.) There are a number of possible explanations for this statement but I shall deal with this more fully in the next section regarding what Plato considered to be the Atlantic.

It also is sometimes suggested that, according to Aristotle, the Pillars of Heracles were also known by the earlier name of 'Pillars of Briareus'. This originates from a statement by Aelian (a Roman writer and rhetorician who lived *circa* 175 – 235 CE) in his work *Various Histories.* (No writings of Aristotle have been found in which he states this.) In chapter 3 of

Aelian's fifth book he wrote *"Aristotle affirms that those Pillars which are now called of Hercules, were first called the Pillars of Briareus; but after that Hercules had cleared the Sea and Land, and beyond all question shewed much kindness to men, they in honour of him, not esteeming the memory of Briareus, called them Heraclean."*[241] This was written nearly 500 years after Aristotle's death, so it is of questionable veracity, and also (and even if this statement is correct) this does not take our issue (as to what Plato intended to convey as the position of the Pillars) any further.

Nor does this affect the, then, generally accepted view that the Pillars were at, or near, the Straits of Gibraltar.

5.1.3.2   Pseudo-Aristotle

There is also another work, *De Mundo,* which was originally attributed to Aristotle but which is now, generally, attributed to a so-called *Pseudo-Aristotle.* It is uncertain when this work was written but the proposed date is somewhere between 350 and 200 BCE.[242] Given that this work does not originate from Aristotle himself, there must be grave doubts about its value in determining what Plato regarded as the Pillars of Heracles.

That said, *De Mundo* does mention the Pillars of Heracles and the Atlantic. At 393a it[243] reads *"Again, the sea which lies outside the inhabited world is called the Atlantic, or Ocean, flowing around us. Opening in a narrow passage, towards the West, at the so-called Pillars of Heracles, the Ocean forms a current into the inner sea, as into a harbour; then gradually expanding it spreads out, embracing great bays adjoining one another, opening into other seas by narrow straits and the widening out again."* The author also writes, at 393b, that *"Libya extends from the Arabian Isthmus to the Pillars of Heracles, though some describe it as extending from the Nile to the Pillars ...".*

It therefore appears that even the Pseudo-Aristotle, whoever he was, regarded the Pillars of Heracles as being the 'entrance' to the Mediterranean Sea and at the Strait of Gibraltar.

5.1.3.3   Dicaearchus, Eratosthenes, Polybius and Strabo

The proposition that the Pillars of Heracles were at, what is now known as, the Strait of Gibraltar is also supported by the statement, by Strabo[244] (when referring to Polybius' *Histories*), that *"Dicaearchus, Eratosthenes, and Polybius and most Greeks place the Pillars at the Straits".*

'*Dicaearchus, Eratosthenes, and Polybius*' were all famous, learned Greeks who lived around the time of, or just after, Plato. Dicaearchus of Messana was a Greek philosopher, cartographer, geographer and mathematician who lived from *circa.* 350 to 285 BCE, Eratosthenes was an ancient Greek mathematician and geographer with lived from *circa.* 276 to 195 BCE, and Polybius was a Greek historian who lived from *circa.* 200 to 118 BCE.[245]

5.1.4   *Plato's position of the Pillars; a conclusion*
Considering all the above it appears probable that Plato would have considered the Pillars to be at, or synonymous with, the Strait of Gibraltar.

5.2   The Atlantic

5.2.1   *The modern concept of the Atlantic*
The Atlantic Ocean, as defined by modern geography, is the sea that is bounded on the west by the Americas and on the east by Europe and Africa. The word ocean, which we now take to be a reference to either the entire sea surrounding the continents or a large sub-division of it (such as the Pacific or Atlantic oceans), is derived from the Greek word 'Okeanos'.

5.2.2   *The 'Okeanos'*
Okeanos,[246] or the 'ocean river', is first mentioned in Homer's *Iliad. It was* a term employed by many ancient writers to refer to the water, or ocean, which they believed encircled the then known world (the Oikumene/ Oecumene). Apart from being that large body of water it was regarded as the source of all rivers and water. This additional element of the concept was known by Plato as he specifically mentions, and describes, the 'ocean river' in his *Phaeton.*[247]

This is not the meaning in modern English and it is apparent that although ancient geographical names may have come down into modern language they do not always have the same specific meanings and connotations as the original words had.

5.2.3   *What was Plato's concept of the Atlantic?*

What we now have to do (whilst recognising that words and concepts have changed meanings and connotations over time) is to attempt to obtain a proper, contextual, understanding of to what Plato was referring when <u>he</u> referred to the Atlantic.

Phrased another way, we must attempt to ascertain what Plato envisaged when he used the term the 'Atlantic'.

Using the same rationale as was discussed in the section on the Pillars of Heracles, it should be accepted that Plato would have used the word 'Atlantic' to describe whatever sea was generally accepted to be such by people likely to be his audience. To have used the name to describe some other sea would have simply confused his audience and, thereby, detracted from his philosophical message.

In reaching a conclusion we must also bear in mind the issue of the Pillars of Heracles (dealt with in Section 5.1) and the important distinction which Plato makes between what was 'inside' and what was 'outside' the Pillars of Heracles. For Plato, Atlantis was on the outside of the Pillars.

5.2.3.1   Where, or what, was the Atlantic, as then commonly accepted?

The first recorded mention of the name Atlantic[248,249] appears to have been by Stesichorus, an ancient Greek lyric poet who lived between, about, 630 and 550 BCE.[250] Only fragments of his writing exist recorded in the works of other, subsequent, writers. He, like Plato, appears to have used the Greek words "*Atlantikoi pelágei*" (Ἀτλαντικῷ πελάγει) to describe the Atlantic.

Also during the sixth century BCE, Euthymenes of Massalia (or Marseilles as it is now know) described himself as having gone on a voyage in the Atlantic Sea. This voyage is referred to by Seneca the Younger in his *Naturales quaestiones*[251] (written about 65 CE). There is a reference to the Nile apparently flowing into the Atlantic which seems anomalous but it must be recognised that even Herodotus considered it possible that the Nile also flowed into the Atlantic.[252] It has been proposed that the reference to the Nile was, in fact, a reference to the river Senegal.

The next existing, recorded, use of the name Atlantic seems to be by Herodotus (*circa* 484 to 425 BCE). This was years before the *Timaeus* was composed by Plato and over a century after Solon's famous visit to Egypt. Herodotus used the name to describe the seas beyond the Pillars of Heracles (*Histories* I. 202). He, however, used the Greek words *Atlantis thalassa* (Ἀτλαντὶς θάλασσα) rather than "*Atlantikoi pelágei*". The Pillars of Heracles were, as I have shown in the previous section, synonymous with what is now referred to as the Straits of Gibraltar.

Herodotus also mentions the Okeanos but, apparently, questioned the truth of what was said about it. He stated, in his *Histories*, "*As for Ocean, the Greeks say that it flows around the whole world from where the sun rises, but they cannot prove that this is so.*"[253] and, further on, "*And I laugh to see how many have before now drawn maps of the world, not one of them reasonably; for they draw the world as round as if fashioned by compasses, encircled by the Ocean river ...*".[254]

Eratosthenes, an ancient Greek mathematician and geographer (*circa* 276 BCE to 195 BCE) who is said to have calculated the circumference of the earth,[255] also referred to the Atlantic. He, however, saw the Atlantic as larger than we accept it today and saw it as stretching from its current position almost all the way up the east coast of Africa. A reconstructed map of the world as Erastosthenes perceived it can be found on Wikipedia.[256] Eratosthenes' perception is similar to Herodotus' - given Herodotus' recognition that the Erythraean Sea (now, approximately, the Indian Ocean) and the Atlantic were one and the same.

Others such as Diodorus (3.38), (who lived in the first century BCE and, thus about 200 years after Plato) appears to have applied the term 'Atlantic'[257] to describe a part of what is now referred to as the Indian Ocean. This may be, not so much as a result of confusion, but more a recognition of the interconnectedness of all the sea and is similar to how Eratosthenes perceived it.

Other recorded journeys (prior to, or around, the time of Plato) into what is now known as the Atlantic (although not specifically mentioning the word Atlantic) are those of Colaeus[258] (around 640 BCE) and Pytheas of Massalia[259] (around 325 BCE). Colaeus,[260] according to Herodotus,[261] travelled (or been driven by bad weather) beyond the Pillars of Heracles and into the Atlantic. He travelled to Tartessus,[262] a city and area on the Atlantic coast of Spain. Pytheas voyaged to North Western Europe. His description of the journey has not survived, but fragments of it have been recorded in other writers' works. On his voyage he circumnavigated and visited a considerable part of Great Britain and is the first person recorded as having described the Midnight Sun.

Finally, concerning the position of the Atlantic, there is the work, *De Mundo*, referred to in paragraph 5.1.3.2. It, I repeat, reads "*Again, the sea which lies outside the inhabited world is called the Atlantic, or Ocean, flowing around us. Opening in a narrow passage, towards the West, at the so-called Pillars of Heracles, the Ocean forms a current into the inner sea, as into a harbour; then gradually expanding it spreads out, embracing great bays adjoining one another, opening into other seas by narrow straits and the widening out again.*" It therefore appears that the author, whoever he was, regarded the Ocean and the Atlantic as synonymous and that it was the sea lying outside the inhabited, circum-Mediterranean, world.

5.2.3.2   Plato's terminology

Plato, when describing the Atlantic, uses the Greek terms '*pelagos*' and '*thalassa*' and does not use the term *Okeanos*. This appears to be the basis of George Sarantitis'[263] disagreement with the translation of Plato's words as a reference to the "Atlantic <u>Ocean</u>". This, in my opinion, is a classic example of the change of the meaning of words.

Pelagos is generally translated as a reference to the 'open sea' or the 'deep' but Sarantitis makes the point that the word '*pelagos*' is, in Greek, generally applied to a sub-area of a greater whole. Hence one finds the Ionian sea, the Aegean sea, the Dalmatian sea and others (all part of the greater Mediterranean Sea) using, in Greek, the word '*pelagos*' - as distinct from other Greek words signifying 'sea' (e.g. *pontus*). The Greek Wikipedia also lists a number of these seas[264] and, similarly, says that a '*pelagos*' is a sea which is less than that of the sea of which it is part.

Sarantitis also makes the point that there is a significant distinction between the concepts embodied by the Greek words "*pelagos*" and '*Okeanos*' and that Plato was aware of this distinction. This is correct. The *Okeanos*, or Ocean River, had a very specific meaning. It was primarily a mythical concept which referred to the waters (ocean, sea or river) believed to surround the habitable parts of the earth and which was the source of all rivers. Plato was definitely aware of the concept of the *Okeanos* (he describes this in his *Phaeton*[265]) and the fact that he deliberately chose to not use *Okeanos* could be indicative of the fact that he did not want to use such a mythical concept with all the connotations which came with that. The contention deliberately used the words '*pelagos*' and '*thalassa*'[266] to avoid confusion with the earlier, and more mythical, concept of the *Okeanos* may be supported by Plato's use of the term '*pan-pelagos*'[267] in the *Critias*[268] when referring to the greater seas (or ocean) beyond Atlantis. ('*Pan*' is a Greek word, or prefix, conveying the concept of "all", "every", "the whole" and/or "all-inclusive"[269].)

Herodotus who, it must be remembered, preceded Plato recognised the 'non-scientific' nature of the concept of *Okeanos*. He wrote that "... *I for my part know of no river Ocean existing, but I think that Homer or one of the poets who were before him invented the name and introduced it into his verse*"[270] and also "*As to the Ocean, they say indeed that it flows round the whole earth beginning from the place of the sunrising, but they do not prove this by facts.*"[271]

If, however I am wrong and Plato did accept the existence of the *Okeanos* then it appears to me that he would still have been correct, (assuming that he wanted to refer to a specific, lesser, part of the *Okeanos* around Atlantis) to have used the word '*pelagos*' to refer to that. The Atlantic (as was recognised by Herodotus) is merely a part of the greater body of the ocean, and even more so the area around Atlantis, so the use of the term '*pelagos*' would, in that sense, be correct.

It is, thus, apparent that Plato's use of the Greek words *pelagos* and *thalassa* to describe what is now known as the Atlantic or North Atlantic was correct.

5.2.3.3   The 'unnavigable' Atlantic

Some authors, trying to establish that Plato's Atlantic was not the Atlantic as Herodotus and we know it, have proposed that because Plato said that "... *for in those days the Atlantic was navigable ...*" he cannot have been referring to the Atlantic as we know it. In dealing with this aspect one must, again, look at what Plato himself wrote.

He wrote about the mud and the consequent impassibility in both the *Critias* and the *Timaeus*.

In the Jowett translation of the *Timaeus* it states that "... *for in those days the Atlantic was navigable*". The implication of this is that it was not navigable at the time of Plato. That is a reasonable interpretation when those words are considered alone however, a few lines after that, it is also stated that "... *and the island of Atlantis in like manner disappeared in the depths of the sea. For which reason the* sea in those parts *is impassable and impenetrable, because there is a shoal of mud in the way; and this was caused by the subsidence of the island.*"

In the Lamb translation the same paragraphs are translated as "... *For the ocean* there *was at that time navigable*" and "... *and the island of Atlantis in like manner was swallowed up by the sea and vanished; wherefore also* the ocean at that spot *has now become impassable and unsearchable, being blocked up by the shoal mud which the island created as it settled down.*"

In the Jowett translation of the *Critias* it is stated that Atlantis "... *when afterwards sunk by an earthquake, became an impassable barrier of mud to voyagers sailing from hence to any part of the ocean.*" In the Lamb translation the same passage is translated as reading that Atlantis "... *now lies sunk by earthquakes and has created a barrier of impassable mud.*"

It is therefore apparent that, according to Plato, it was not the whole Atlantic Ocean which was impassable or unnavigable but only the area in the vicinity of Atlantis.

It is also apparent, therefore, that the proposal that Plato stated that the whole of the Atlantic was made impassable is not correct. This also leaves open the possibility that the Atlantic referred to by Plato is the Atlantic as we know it today.

5.2.3.3.1   *Aristotle's shallow, muddy sea*

Although it is clear from the above that Plato himself was not referring to the whole Atlantic as being impassable, various people have, nevertheless, sought support for the proposition that it was the whole of the Atlantic in the statement by Aristotle (who came after Plato) that outside the Pillars of Heracles the sea which was shallow and muddy. This is not a description of the vast Atlantic, as we know it; it being neither shallow, calm nor lying in a hollow. (Nor is it a description of the western Mediterranean which is neither shallow nor in a hollow.)

People seeking confirmation of an assumption that Plato's 'Atlantic' is not the Atlantic as we know it tend to accept that Aristotle's reference to the sea "*outside the Pillars of Herakles*" is a reference to the whole of the

Atlantic. Then, knowing that this description does not fit the Atlantic, follow through with the conclusion that Plato could not be referring to the Atlantic as Herodotus knew it and we, to this day, know it. Following from that they conclude that references by Plato to the Atlantic must be references to some other sea.

This convoluted logic ignores the fact that Aristotle came after Plato and that there were others, who had preceded Plato, who understood a reference to the Atlantic to be a reference to the Atlantic as we, generally, now know it.

Nevertheless we can, and should, examine Aristotle's statement to see if there is any other logical explanation for his statement.

To reiterate, Aristotle states, in his work *Meteorology*,[272,273] that *"... outside the Pillars of Heracles the sea is shallow owing to the mud, but calm, for it lies in a hollow."*

This, ostensibly, anomalous statement may be an insertion by a later 'scribe' but, assuming that it is not, there appear to be four possible explanations for it:

1. Aristotle did not understand the Pillars to be at the Straits of Gibraltar, and was thus referring to a sea other than the Atlantic as it is now known,

2. he made a mistake,

3. he, unwittingly, repeated Phoenician / Carthaginian propaganda or

4. the area around the Straits had, at the time that he wrote, become muddy and shallow.

<u>Scenario 1</u>.   Given the history of the Pillars being at the straits of Gibraltar [as has already been discussed in the previous sub-section], given the fact that Aristotle consistently used the term 'The Pillars' to apply to the western extremity of the Greek world and given the fact that by the time of Herodotus [about 100 years before Plato himself] there was knowledge of the sea outside the Pillars[274,275]) it appears unlikely and

implausible that Aristotle would not have been referring to the Straits of Gibraltar and to the Atlantic as we know it.

This scenario, namely that Aristotle did not understand the Pillars to be at the Straits of Gibraltar, and was thus referring to a sea other than the Atlantic as it is now known, therefore appears to be implausible.

Scenario 2.   We all make mistakes and have to rely upon information from others however the scenario that Aristotle simply made a mistake also seems unlikely.

Scenario 3.   The Phoenicians (and subsequently their colony Carthage in north Africa) were great seamen, maritime traders and conquerors and had control over the Western Mediterranean before the Greeks ventured that far west. The Phoenicians ventured beyond the Pillars of Heracles[276] (as is evidenced by Hanno the Carthaginian's periplus) and are believed to have been the first to have reached Britain (which they may have referred to as the Cassiterides[277]). They were also known to have jealously guarded their territory and commercial interests, by the withholding of information about it, propaganda and, if necessary by force of arms.[278] Strabo[279] reported that *"Formerly the Phœnicians alone carried on this traffic from Gades, concealing the passage from every one; and when the Romans followed a certain ship-master, that they also might find the market, the shipmaster of jealousy purposely ran his vessel upon a shoal, leading on those who followed him into the same destructive disaster; he himself escaped by means of a fragment of the ship, and received from the state the value of the cargo he had lost."*

It is therefore quite possible that the statement reflects Phoenician / Carthaginian propaganda spread to discourage potential competitors and, as a result, Aristotle was simply repeating this 'propaganda' which had become generally accepted.

Scenario 4.   The final possible scenario is that area around the Strait had, at the time that Aristotle wrote, in fact become muddy and shallow. Is there anything to corroborate such a contention?

At present, the greater Western Basin of the Mediterranean Sea varies from about 3 000 to about 1 000 metres deep, and the Atlantic (beyond the shallow mouth of the Straits of Gibraltar) also drops to about 3 000 metres deep. Just outside the mouth of the Mediterranean in the relatively shallow Atlantic mouth there is an area known as Spartel Bank (35.916667°N 5.966667°W).[280] This area, which is now submerged, was previously an island[281] but is only about 58 metres deep. This is, in relation to the greater Atlantic, very shallow. It would have been even shallower around the time of Atlantis[282] and also more recently.

The whole general area is prone to regular (millennial scale) powerful earthquakes. Gutscher wrote[283] that *"Indeed, the regional paleoseismic record ... suggests great earthquakes off SW Iberia every 1500-2000 years. Tsunami deposits indicate an earlier great earthquake struck SW Iberia around 200 BC, as noted by Roman records from Cadiz."*

What is significant about this quotation is that date of the earthquake is "<u>around</u> 200 BC". Is it possible that it (or another powerful precursor) actually occurred within the lifetime of Aristotle?

One does not know how accurate the stated date of the earthquake is, whether there were any other precursors to it or, for that matter (and inasmuch as Aristotle referred to it as lying in a hollow), whether Aristotle was referring to a specific part, bay or estuary. Cadiz is situate in a bay and had an estuary.

The prevalence of earthquakes in this area, offshore of the Gulf of Cadiz, and the earthquake around this time are also referred to in an article by P G Silva *et al* entitled *"Seismic palaeogeography of coastal zones in the Iberian Peninsula: Understanding ancient and historic earthquakes in Spain."*[284]

Given these uncertainties and possibilities, it is questionable whether much weight can be given to Aristotle's statement. Nevertheless, it appears possible that areas around the Straits could have been affected by earthquakes and could, also have become *"muddy and shallow"*, thereby matching Aristotle's description.

5.2.4   *Plato's Atlantic; a conclusion*

In conclusion it is, therefore, probable that Plato was referring to the Atlantic as we now know it when he wrote of the Atlantic. At the very least he would have been referring to that portion of the Atlantic, as we know it, north of the equator.

5.3 Continent

5.3.1 *What is meant by English term 'continent'?*

The term 'Continent', is derived from the Latin *terra continens*, meaning 'continuous land' and, therefore, did not form part of Plato's vocabulary. Latin, the Romans' language, only became ascendant with the waning of Greek power and influence and even in English the word is relatively new, having not come into use until the Middle English period (11th -16th centuries).[285,286]

Wiktionary defines 'continent' as "*A large contiguous landmass considered independent of its islands, peninsulas etc.*"[287] whilst in Wikipedia it is stated that "*A continent is one of several very large landmasses of the world. Generally identified by convention rather than any strict criteria, …*".[288]

5.3.2 *Ancient Greek concept of 'continent'*

Ancient Greeks did not, in the early days at least, have a specific word having the same specific meaning as the English word 'continent'.[289] By way of example, Herodotus at 4.118.1[290] in his original, Greek language, *Histories*[291] uses the word πείρ/ipeiro. This word has been translated as 'continent' (referring to Asia and Europe) but it, more literally, means simply "terra firma" or "land".[292]

Prior to Herodotus only Europe and Asia were considered continents, with 'Libya' sometimes considered part of Asia. Herodotus, who wrote after Solon but before Plato, accepted that there were three known continents namely, Europe, Asia and Libya [Hdt. 4.42]. That said he queried the logicality of this division. In Book 4 of his *Histories* (as translated by G Rawlinson[293]) Herodotus wrote "*For my part I cannot conceive why three names, and women's names especially, should ever have been given to a tract which is in reality one, nor why the Egyptian Nile and the Colchian Phasis (or according to others the Maeotic Tanais and Cimmerian ferry) should have been fixed upon for the boundary lines; nor can I even say who gave the three tracts their names, or whence they took the epithets.*" A little further on he, nevertheless, writes "*However let us quit these matters. We shall ourselves continue to use the names which custom sanctions.*"[294]

It is apparent, from this last statement, that Herodotus was bowing to custom when accepting the division into three continents.

It is also apparent that, by the time of Herodotus, the Ancient Greeks (even though they did not have the identical word 'continent') accepted that the known land mass was notionally divided up into various sub areas.

In everyday English the term 'continent' can also be 'fuzzy' and relative. It is sometimes used to imply a mainland.[295] In this vein, the main Scottish island of the Shetlands is known as Mainland. It is suggested that this is so because, amongst the Shetland isles, it is the main island. Similarly, the inhabitants of the island Mainland may apply the same designation to the rest of Britain, who in turn apply the term continent or 'continuous land' to mainland Europe.[296]

This fuzzy terminology may also affect translations. In the Yonge translation of Philo of Alexandria's (20 BCE-50 CE) *On the Eternity of the World* a question is put *"Are you ignorant of the celebrated account which is given of that most sacred Sicilian strait, which in old times joined Sicily to the continent of Italy?"* (v.139)[297]. Colson translates this same question, and related section, as *"Do you not know the celebrated story of the sacred Sicilian straits? In old days Sicily joined on to the mainland of Italy ..."*.[298] In the one translation Italy is referred to as a continent but in the other it is referred to simply (and more in accordance with modern English) as the mainland.

As an aside, this description is probably a reference to the formation of the Strait of Messina referred to in section 5.1.1.

There are many examples of the continuing usage of the word 'continent' (or even the comparable word in other languages) in a relative, or non-scientific, sense. This 'fuzziness' compounds the problem when translating from, in this case, Ancient Greek into English.

It is also possibly important to understand that it is not only 'fuzzy' terminology that may cause problems in translation; names (or what they designate) may also change. The term 'Italy' was originally used in pre-Roman times to describe only the southern part of the Italian Peninsula, not the entire peninsula as it now does.[299]

5.3.3   *What did Plato intend to convey by the word translated as 'continent'*

What is important is, as always, to attempt to ascertain what Plato intended to convey by the words he used.

In the Atlantis story the translated term 'continent' is first used in *Timaeus* 24e. The relevant portions of sections 24e to 25b, which both mention 'continent', have been translated as "*... it was possible for the travellers of that time to cross ... to the whole of the continent* [25a] *... which encompasses that veritable ocean. For all that we have here, lying within the mouth of which we speak, is evidently a haven having a narrow entrance; but that yonder is a real ocean, and the land surrounding it may most rightly be called, in the fullest and truest sense, a continent. Now in this island of Atlantis ... which held sway over all the island, and over ... parts of the continent; and, moreover,* [25b] *of the lands here within the Straits they ruled over Libya as far as Egypt, and over Europe as far as Tuscany...* ".[300]

In the Jowett and Lamb translations of the latter part of the above quote this is translated as, respectively, that the Atlanteans "*had rule over ... parts of the continent, and, furthermore, ... had subjected the parts of Libya within the columns of Heracles as far as Egypt, and of Europe as far as Tyrrhenia.*"[301] and that the Atlanteans "*held sway over ... parts of the continent; and, moreover... of the lands here within the Straits they ruled over Libya as far as Egypt, and over Europe as far as Tuscany.*"[302]

The original, Greek, word which was translated into 'continent' (as the best possible translation) in the last line of this quote is "ἤπείρου". This did not, originally, mean continent and may, more literally, be translated as "to make into a mainland".[303]

That said, what is important is that Plato makes a clear distinction [in *Timaeus* 25a and 25b] between the "*continent*", on the one hand, and "*Libya .. and ... Europe*", on the other. For him, they are separate and distinct land areas.

It follows, by necessary implication (Europe and Libya being inside 'the Columns of Heracles / Straits'), that the continent must be outside the Columns / Straits, which, as we have seen, referred to the Straits of Gibraltar.

Plato, in *Timaeus* 25a, also states that the land mass in question may (depending upon the translation) "*be most truly*", "*most rightly be*" or is "*correctly*" called a continent. This "*truly*", "*rightly*" and "*correctly*" would, if this was a reference to Europe, have been redundant. It was known that Europe and Asia (and Libya) were continents and it would not have been necessary to state that the land in question "*may*" "*truly*" or "*rightly*" be called a continent. The reference to "continent" must, therefore, be a reference to some large land mass which was either not generally known or was not generally accepted as being a continent. Insofar as it has been suggested that this 'continent' couldn't have been outside the known, Mediterranean sea, the Ancient Greeks were (as we know from Herodotus) aware of the sea outside the Mediterranean. He was also aware, as also appears from the *Histories*, that Africa had been circumnavigated and that one could sail by sea from the Erythraean Sea to the Atlantic.[304] Modern geology has definitively demonstrated that no continental mass lies in the Atlantic and quite clearly the Mediterranean does not have room for a sunken continent.

As an aside, Claudius Aelianus, in his work *Various History*, quotes Theopompus (a Greek historian who lived between 380 BCE and circa 315 BCE) as writing that "*that Europe, Asia and Africk were Islands surrounded by the Ocean: That there was but one Continent[305] onely, which was beyond this world, and that as to magnitude it was infinite.*"[306] It is unclear what, exactly, Theopompus meant by the statement that Europe, Asia and "Africk" were islands but what is apparent and unambiguous is that he recognised that there is a one continent which is separate from Europe, Asia and Africa.

5.3.4   Continent, a conclusion

A number of commentators have assumed that when Plato referred to an 'opposite continent' he was referring to the Americas. The conclusion that this is a reference to the Americas is, on the simple wording, possible but is not supported by archaeological, or other, studies.

5.4 The size of Atlantis – 'larger than Libya and Asia combined'

The question of the size of Atlantis is also dealt with in Section 7 where I consider the measurements given by Plato. Here, however, I shall analyse the specific statement that Atlantis was larger than Libya and Asia combined and also some of the terms and descriptive words used by Plato.

We must, again, bear in mind, firstly, that we are attempting to ascertain what Plato intended to convey as the sizes of 'Libya' and 'Asia' and, secondly, that he would not have used terms which were unknown to his audience.

Before dealing with these issues it needs to be remembered that Atlantis was not continent sized. This is an erroneous conclusion based upon the assumption that, when Plato referred to Libya and Asia, his concepts of what these names designated were the same as we apply to them today. Contrary to this Plato unambiguously and only referred to Atlantis as an island. (Tony O'Connell at Atlantipedia has, usefully, listed all Plato's references to Atlantis as an island.[307])

Plato would have assumed that his audience knew to what he was referring when he referred to Libya and Asia and, as such, he does not give any indication of the size of these areas. It may, nevertheless, be possible to extrapolate from statements by, at least, Herodotus as ascertain the size of Plato's Libya and Asia.

5.4.1 *Meizon*

Before entering into the issue of the sizes of Libya and Asia (which is fraught with difficulty) it is necessary to look at the word that Plato uses and which has been translated as "larger" in the Jowett, Perseus Tufts / Lamb and Bury translations. This is the word "*Meizon*".

Meizon has been translated consistently, across the Jowett, Lamb (Perseus Tufts') and Bury translations as 'larger' however some 'Atlantis investigators' have suggested that the word, where used in *Timaeus* 24e, should be translated as 'greater'[308] - rather than 'larger'. The concomitant proposal by these investigators is that Plato, when using the word, sought to convey that Atlantis was 'greater in power' and not 'larger' than Libya and Asia.

Accepting that *meizon* may be translated as 'greater' and, even (for the sake of this discussion) that that may be the predominant translation, does not mean that it is the only correct translation. According to the various lexicons at the Perseus Tufts website[309] *meizon* has been translated as, amongst others, "*vast, high wide and long*"[310] and "*vast, spacious, wide*"[311].

Obviously, how the word should be translated in any particular situation depends upon the context in which it is used.

The relevant portion reads "*for in front of the mouth which you Greeks call, as you say, 'the Pillars of Heracles' there lay an island which was larger (μεἰζων/meizon) than Libya and Asia together; and it was possible for the travellers of that time to cross from it to the other islands, and from the islands to the whole of the continent*"

The word, *meizon*, as used here, describes the island of Atlantis and, in particular, Atlantis in the context of its position, travel and, by implication, distances. Whilst it is notionally possible that *meizon* could mean 'more powerful', or in some way 'greater', there is nothing in the context to suggest this. There is no reference, in the particular context, to the might, power nor grandeur of Atlantis. If the reference was intended to be to the warlike or controlling power of Atlantis there would, one would expect, be some indication of the travel being controlled or conducted by the powers of Atlantis. This is not apparent and, to the contrary, there is an implication of freedom of movement.

I, accordingly, doubt that it was Plato's intention to convey the power of Atlantis and agree with Jowett's, Lamb's and Bury's translations – namely that *meizon* is a reference to the size of Atlantis and could, as such, be legitimately translated as 'larger' in area. That said even the word 'greater' in English can have various connotations depending on the context in which it is used and can also mean 'larger'.

Be that as it may, and even if it is a reference to the power of Atlantis, that does not detract from the size of Atlantis as estimated from the size of the central plain and other descriptions dealt with in Section 7.

5.4.2   'Libya and Asia combined'

It is extremely difficult, given that we do not know what Plato considered to be Libya and/or Asia, to make any dogmatic statement about this aspect of the description. Plato, as I said above, does not define what is meant by these terms and simply accepts that his audience knows what they describe. The concept of what 'Asia' embraced had, in ancient times, a variety of meanings[312] but what is never was, was the vast continent that it is today. Similarly, there is uncertainty as to what was embodied in the name 'Libya'[313] but it could not, logically, have referred to the whole of Africa. It was, generally, taken to refer to North Africa or a portion of that.

Nevertheless and given Herodotus' statement that "*Europe* [was] *in length equal to Asia and Libya combined*"[314] we can extrapolate and obtain some idea of the size intended to be conveyed by Plato's statement. Herodotus also, conveniently, gave a description as from where Europe was regarded as starting. He said the boundary between Europe and Asia was "*the Colchian Phasis river*" or (as he acknowledged that this was not accepted by all) the *Cimmerian Ferry*. The 'Colchian Phasis river' is what is today known as the Rioni River[315] in the country Georgia whilst the area of the 'Cimmerian Ferry' is generally regarded as being the strait between the Black and Azov seas, known as the Kerch Strait[316].

We therefore know from where, approximately, the measurement started but to where it was measured we do not know. Logically it must however it must have been somewhere on the Atlantic shore. Accepting that Europe must have ended where it reaches the Atlantic I have considered two destination points, Brest and Calais in France.

The distance, by foot (according to Google maps), from the "Colchian Phasis river" to Brest in France (possibly the western-most point save for the Iberian Peninsula) is approximately 4 100 km's[317] but to Calais is about 3 500 km's[318]. The distances from the Kerch Strait to Brest and Calais are shorter and are, respectively, about 3 540[319] and 2 950 km's.[320]

5.4.2.1   The length of Asia and Libya

Given those measurements, the length of Herodotus' (and, presumably, Plato's) Europe was between, about, 2900 kilometres and 4100 kilometres. We therefore know, from Herodotus' statement, that the length of

"Asia and Libya combined" was, similarly, between 2900 kilometres and 4100 kilometres.

The variation in distances, about 1200 kilometres, is significant but at least it gives us an approximate distance for the length of Europe and, consequently, of the size of Atlantis. We also therefore know Atlantis was not (apart from the statements that it was an island) a truly massive continent.

5.4.2.2   The measurement of Atlantis

Whilst we have an approximate measurement we still do not know precisely what was contemplated by Plato when he said Atlantis was larger than Libya and Asia combined. To what measurement was he referring? Was it surface area, circumference or length?

Felice Vinci in his book, *The Baltic Origins of Homer's Epic Tales*[321], states that the ancient Greeks typically measured the size of an island by its coastal circumference - rather than by its surface area, as we tend to do today. He refers, in support of this, to Diodorus Siculus'[322] description of Great Britain[323] and also comments that this method was used by Christopher Columbus on his journeys. Possible further corroboration of this is the fact that Polybius states (albeit critically), in his *Histories,*[324] that according to Pytheas, " *… the coastline of* [the whole of Britain] *is more than forty thousand stades."*

Accepting this method of measurement we then have an idea of the circumference of Atlantis.

5.4.2.3   The measurement of Great Britain and Hatton Rockall

My proposition is that Atlantis / Hatton Rockall would, if it were an island today, be roughly the size of Great Britain. (I deal with this in section 7.2.2.2) Great Britain, actually being an island, is a lot easier to measure so I shall consider, first the maritime perimeter of Great Britain and then the perimeter of Hatton Rockall.

Using Google Earth (and not taking into account every 'nook and cranny' but 'travelling' in straight lines) the distance around Great Britain is about 2860 kilometres. According a calculation done on the website 'ports.com',[325] the distance sailing around Great Britain (sailing from

Cowes to Dundee to Stornoway to Port Ellen to Ramsey to Pembroke to Newquay to Plymouth and back to Cowes) is about 3430 kilometres.

Using Google Earth (and, again, not taking into account every 'nook and cranny' but 'travelling' in straight lines) the distance around Hatton Rockall is about 2730 kilometres. It is probably safe to assume that, similar to the ports.com calculation of the sailing distance around Great Britain, the actual sailing distance is around Hatton Rockall would be larger than that calculated on Google Earth.

Using the same ratio (i.e. between the distances around Great Britain according to Google Earth and according to ports.com) one can calculate that the sailing distance around Hatton Rockall as an island could be approximately 3275 kilometres.

5.4.2.4   Conclusion

I propose, considering that a minimum length[326] of "Libya and Asia combined" was in the order of 2950 kilometres and the circumference of Hatton Rockall is in the order of 3275 kilometres,[327] that the distance around Hatton Rockall is comparable with and probably exceeds the lengths of Plato's Libya and Asia.

This, when taken in conjunction with the other aspects of Atlantis discussed in Section 7, lends credence to the possibility that Atlantis was Hatton Rockall.

SECTION SIX:
WHERE WAS THE ISLAND OF ATLANTIS?

6.1   Reiteration of relevant elements

In the *Timaeus* it is stated that *"there was an island situated in front of the straits which are by you called the Pillars of Heracles"*. In the *Critias* it is simply stated that the Atlanteans were *"those who dwelt outside the Pillars of Heracles"*. This, as we have seen, placed Atlantis in the Atlantic but that still leaves the question as to where, within the Atlantic, it was.

Another potential indicator of the position of Atlantis within the Atlantic is the statement that "[Atlantis] ... *was the way to other islands, and from these you might pass to the whole of the opposite continent ...*".

Plato's description of the environment of Atlantis is one of lush growth and an equable climate. He states that:

> ➢ *"There was an abundance of wood for carpenter's work, and sufficient maintenance for tame and wild animals."*

> ➢ *"... for as there was provision for all other sorts of animals, both for those which live in lakes and marshes and rivers ..."*

> ➢ *"Also whatever fragrant things there now are in the earth, whether roots, or herbage, or woods, or essences which distil from fruit and flower, grew and thrived in that land ..."*

> ➢ *"Twice in the year they gathered the fruits of the earth—in winter having the benefit of the rains of heaven, and in summer the water which the land supplied by introducing streams from the canals."*

> ➢ *" ... also the fruit which admits of cultivation, both the dry sort, which is given us for nourishment and any other which we use for food - we call them all by the common name of pulse, and the fruits*

6.2   Analysis

6.2.1   Atlantis, "in front of" the Pillars of Heracles

It is only in the *Timaeus* that it is stated that Atlantis was "*in front of*" the Pillars of Heracles. The Greek word used by Plato, in the *Timaeus*, to describe the position of Atlantis (in relation to the Pillars of Heracles) was "πρό" or, in the Latin alphabet, "*pro*". The concise, and general, translation of this word is "*before*" or "*in front of*".[328] This is consistent with the above translation. The word can, however, also be interpreted (depending on its context) in other ways including having the meaning "*outside*" (e.g. outside a house) attributed to it.[329,330]

Stephen Kershaw, the author of *A Brief History of Atlantis: Plato's Ideal State*,[331] agrees that it could be translated as "outside" but comments that the word "*pro*" generally has a connotation of proximity.[332]

It is, therefore, possible that the interpretation "*in front of*" might be too narrow and that a better translation might be, simply, that Atlantis was outside the Pillars of Heracles. This would then be consistent with the statement in the *Critias*.

6.2.2   Atlantis, a temperate climate

The description given of Atlantis is, but for the aspect with which I shall deal with below, consistent with a temperate climate. It could, therefore, well have been "*in front of the straits*" of Gibraltar. As my hypothesis places Atlantis north of "*in front of the straits which are by you called the Pillars of Heracles*" the legitimate question which must arise is, 'Is there any basis for looking north of "*in front of the straits*"?'

6.2.2.1   Temperate climates generally

A temperate climate is one in which temperatures are generally relatively moderate (i.e. neither extremely hot nor cold) and in which the changes between summer and winter are also usually moderate.[333] The climate of an area is affected not only by its position North or South of the Equator but also by the overall environment in which it is found, including its proximity to the sea and mountains. (I shall discuss this in more detail in section 8.3.)

The area in the north eastern Atlantic between 30^0 and 60^0 N would generally fall within the definition of a temperate climate. This area covers from the western (or Atlantic) coast of Morocco to just north of Scotland in the United Kingdom. (See Figure 6.1.) [To the contrary the western Atlantic at 60^0 N (approximately the southern tip of Greenland) is far from temperate.] This area (between 30^0 and 60^0 N in the eastern north Atlantic) would include the area "*in front of*" the Pillars of Heracles.

The prevailing winds in this area are westerlies bringing most rain during the winter months.

6.2.2.2   A 'jarring note'

There is one small, but jarring, description given by Plato which allows me to suggest that one may look further north than "*in front of the straits*". It is in his description of the plain.

He says, in the Jowett translation, "*This part of the island looked towards the south, and was* <u>sheltered from the north</u>*.*" [my underlining]. In the Perseus Tufts/Lamb translation this is translated as "*And this region, all along the island, faced towards the South and was* <u>sheltered from the Northern blasts</u>*.*"[334] [again, my underlining].

I understand these "*Northern blasts*", from which the plain was sheltered, to be strong and, presumably, unpleasantly cold northerly winds and, most probably, accompanying inclement weather.

It is also, given their specific mention, reasonable to assume that these "northern blasts" were not an uncommon event – there would have been no necessity to mention being protected from them if it was an infrequent, or unusual, event.

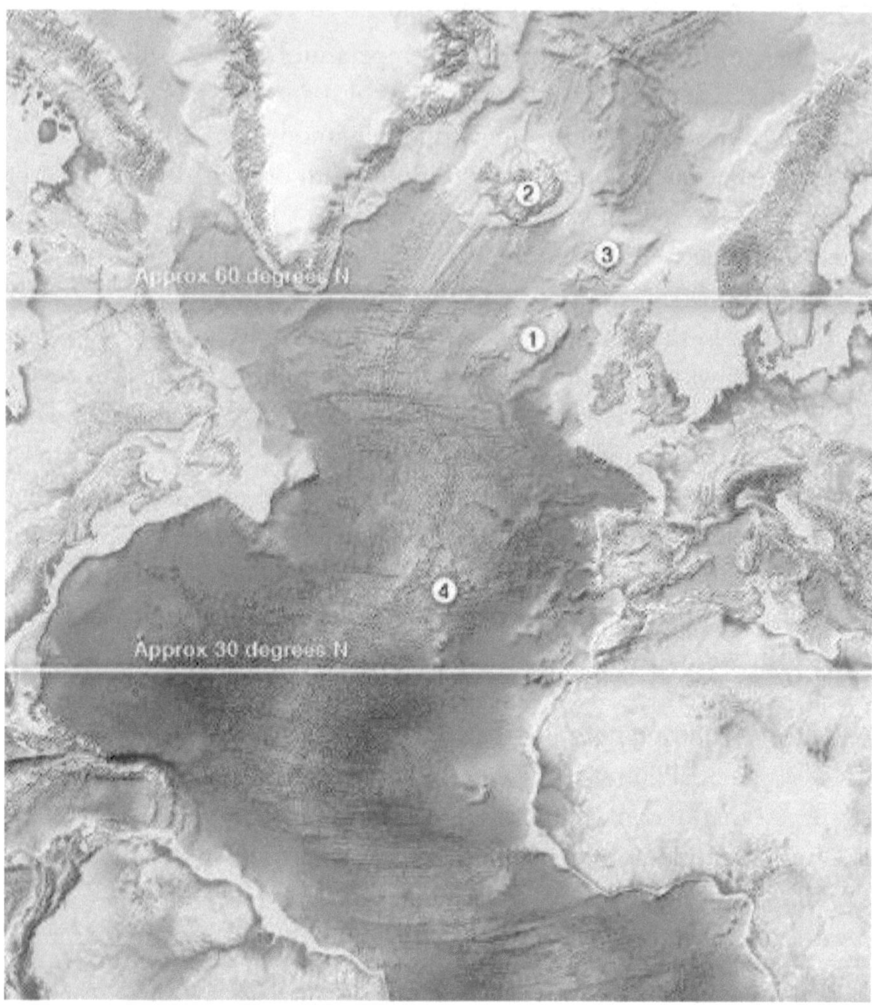

Figure 6 1 Bathymetric map of the northern Atlantic Ocean (showing approx. position of 30°N and 60°N and the approx. positions of (1) Hatton Rockall, (2) Iceland, (3) the Faeroe Islands and (4) the Azores. [335]

6.3 Atlantis, a way to other islands and to the opposite continent

If, as I mention in section 5.3, the reference to the "*opposite continent*" is a reference to, at least, North America then the reference to 'other islands' also becomes potential corroboration of my proposed position of Atlantis.

Sea level was, at the end of the Pleistocene and the beginning of the Holocene, lower than at present and, as will be seen from Figure 7.5, there were probably a number of other islands in the vicinity of Hatton Rockall. (The possibility has also been referred to in section 3.4.2) The distances between these islands themselves (and Hatton Rockall, the Faeroes, Iceland and Greenland) and also to North America were, compared with the greater Atlantic, short. Early sailors could have travelled the relatively short distances from island to island and then to North America, as did[336] the early Norse. These distance would also have been, as will be seen from Figure 6.1, substantially less than those from the Azores to the Americas.

Therefore, for travel by 'island hopping' to the 'opposite continent, it becomes far more likely that Atlantis was at Hatton Rockall than in the Azores.

6.4 Conclusion

Given the cryptic comment about the northern blasts (And the fact that the climate of the Azores is not one that experiences significant cold northern blasts) I am of the opinion that one may look further north than simply "*in front of the straits*".

If it is accepted that one may look further north, then we may start with considering, for completeness sake and although they are not sunken, the Faeroe Islands and Iceland. The Faeroe Islands and Iceland cannot be described as having a temperate climate and are unlikely to have had in the last 12 millennia. Furthermore, they also lie north of 60^0 N - at approximately 62^0 N and 64^0 N respectively.[337]

Considering the actual geography of the north Atlantic, it is clear that the only[338] significantly large submerged area (a potential large island) in the North Atlantic is that of Hatton Rockall (See Figure 6.1).

This, and the corroboration given by the statements about the travel to the *"other islands"* and onward travel to *"the opposite continent"*, makes it reasonable to consider Hatton Rockall as a possible site for Atlantis.

In Section 7 I shall consider the topography of Hatton Rockall and set out the congruities between Hatton Rockall and Atlantis' natural\ environment. In Section 8 I consider possible evidence for the fact that Hatton Rockall's was an island during the Pleistocene and also propose that Hatton Rockall's climate, as an island, might have been similar to that of Ireland's present climate.

SECTION SEVEN:
COMPARISON OF TOPOGRAPHIES OF ATLANTIS AND HATTON ROCKALL

7.1   Introduction

In this section the topography (or, possibly more correctly, bathymetry) of Hatton Rockall is compared with Plato's descriptions of the island of Atlantis.

Before considering the physical structure of Hatton Rockall two general matters may be considered first:

> Firstly, "was Hatton Rockall ever, according to our current understanding of sea level changes and tectonic events, above sea level?" and

> Secondly, is it possible for large areas of land to be subject to repeated subsidence and exhumation?

7.1.1   Was Hatton Rockall ever above the sea?

The Hatton Rockall area, or parts of it, have been described, variously, as a "*continental fragment*",[339] a "*submerged continental crustal block*"[340] and a "*micro continent*".[341, 342] As can been seen from Figure 7.1 it also falls within an area of extended continental crust. This designation as 'continental' implies, of itself, that it was formed above the water. It has also been found that there are "*Large wedges of mainly sub-aerially erupted basalts ...*"[343, 344] 'Subaerial', which literally means "under the air", is used to describe events or features that are formed on the Earth's land surface and exposed to Earth's atmosphere.

The answer to the first of the above questions must, therefore, be 'yes'. The Hatton Rockall area was formed above water and it has, at some later stage, become submerged. The mere fact that it might be accepted that Hatton Rockall was above sea level millions of years ago does not, I recognise, mean that it was 12 000 years ago.

7.1.2   *Can large areas of land sink and rise?*

One simply has to consider the north Atlantic area in this regard. The North Atlantic edges have, on both sides of the Atlantic and over millions of years, been subject to both episodic subsidence and uplift. (This is not restricted to the North Atlantic but has also occurred repeatedly at various locations around the world.) I shall deal with this phenomenon more fully in Section 13.

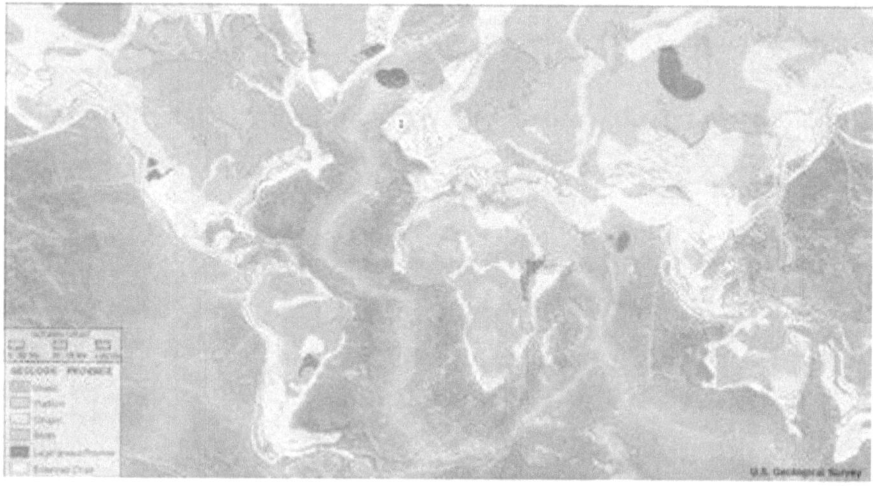

Figure 7 1 World geologic provinces (1. marks the position of Hatton Rockall) [345]

Indicators of the history of an area's subsidence and/or uplift are found in, amongst other things, the sedimentary rocks and sediments which are found in that area.[346] Anomalies, or breaks, in the laying down of sediments may be indicative of a sub-aerial period. It is therefore important to note that there have been fluctuations, and possibly even complete breaks, in the laying down of sediments in the Hatton Rockall area.[347] Whether these fluctuations are attributable to Hatton Rockall being above the surface of the sea or whether they are attributable to fluctuations in the productivity of micro-plankton, ice-rafted input or to other factors[348] is something which still has to be determined with certainty.

ATLANTIS, FOUND?

At least one section of the Hatton Rockall area was above the water sometime during the past, approximately, 12 000 years [349] - dredged material which indicates that the area was above sea level during the Holocene has been recovered from areas which are now in excess of 100 metres deep.[350] Other areas, including on the outer edge of Hatton Bank, have been subject to exposure and weathering,[351] which could only have occurred whilst it was subaerial. Hatton Rockall itself has, furthermore, been shown to have sunk by at least 1400 meters[352] - precisely when this occurred is still uncertain.

The Hatton Rockall area could, it appears, have been above sea level at some stage in recent geological history and, as such, we will continue the investigation. The possibility of Hatton Rockall having been an island and the possible causes for it having sunk to its present depth will be investigated further in Sections 8 and 13.

I shall now move on to set out, and compare, the topographies of Atlantis and Hatton Rockall.

7.2   The size and general topography of Atlantis

7.2.1   *Reiteration*

I reiterate that Plato stated that Atlantis:
  - was a large island situated in the Atlantic

  - had a central plain

  - which looked to the south and

  - was surrounded by mountains which

  - were lofty and

  - descended precipitously to the sea on the side away from the plain and

➢ sheltered the plain from the northerly winds

➢ had in the centre of the island a hill, or small mountain

➢ had natural hot springs and

➢ was the passage to other islands.

7.2.2 Discussion

7.2.2.1 The size of Atlantis

I have already touched on the question of the size of Atlantis in section 5.4 (in relation to the statement that Atlantis was *"larger than Libya and Asia put together"*) but here I shall consider the indicators of size included in Plato's description, particularly the plain, the mountains and its relationship to other islands.

7.2.2.2 The plain

Plato gives us the measurements of the central plain of the island. In the Jowett translation he describes it as *"extending in one direction three thousand stadia, but across the centre inland it was two thousand stadia"*. The *Perseus Tufts* translation (par. 118a) states *"this plain had a level surface and was as a whole rectangular in shape, being 3000 stades long on either side and 2000 stades wide at its centre"*.

Hatton Rockall has (as can be seen from, particularly, Figure 7.2) a central plain, which is largely aligned North/South.

There is no certainty as to how long the stade referred to by Plato was but it is generally accepted that a stade varied, depending upon the culture in which was being used, between 157 and 209 metres in length.[353] The measurements of the plain would, therefore, have been (depending upon which length of the stade is used) between 471 kilometres by 314 kilometres, at its smallest, or 627 kilometres by 418 kilometres at its largest. (I shall deal with the specific measurements of the Hatton Rockall plain in Section 7.3.)

As an aside, and for those who favour a continental sized Atlantis, this would suggest a super-continent (the combined size of Asia and, at least northern, Africa - as we now know them) comprised entirely of mountains save for a single (relatively speaking) very small central plain. There is no indication of such an unusual topography in the story. As such it becomes more plausible that Atlantis was (apart from Plato's multiple statements to that effect) simply a large island.

The simplest way, possibly, in which to visualise Hatton Rockall is to provide a map of the area as if it was an island. Figures 7.2 and 7.3 are such diagrams, with north being at the top of the diagrams. Figure 7.3 is a map showing the relative size and position of Hatton Rockall if it were an island. Further diagrams, showing the profile of the plain and surrounding high areas, can be found in the Appendix.

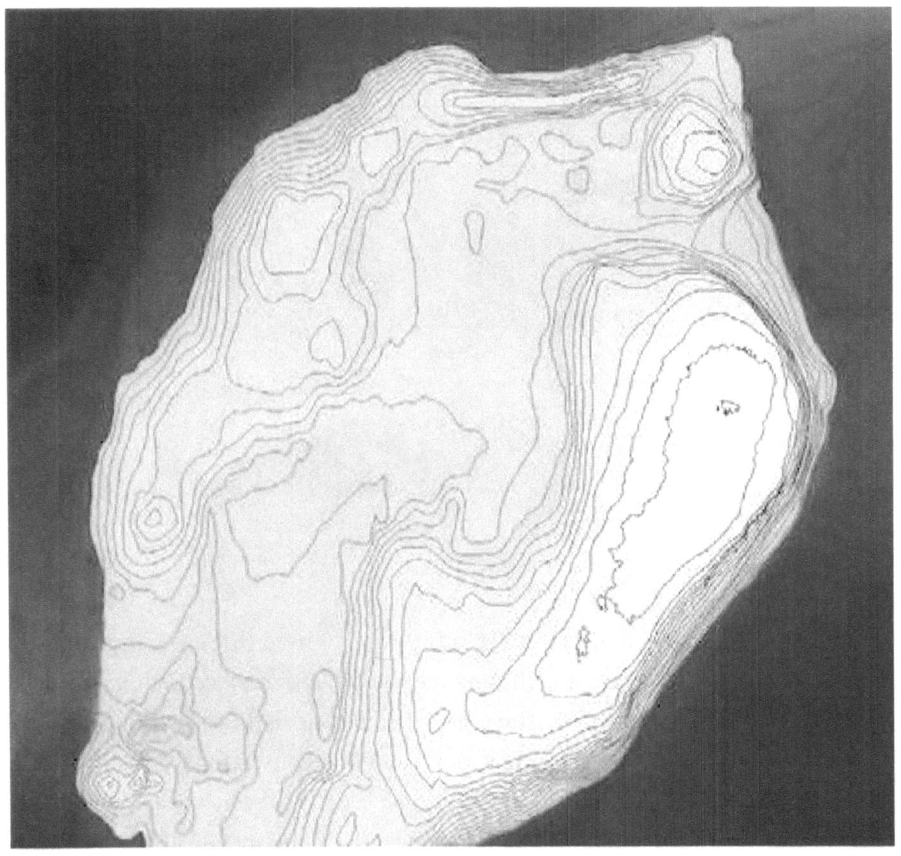

Figure 7 2 Hatton Rockall as an island [354]

Hatton Rockall measures, at a depth of 1 000 fathoms (or 1 828 metres), approximately 740 kilometres long by 445 kilometres wide and, at this depth, it has a 'surface' area of 220 000 km². [355] This would make Hatton Rockall, as an island, a large island and, probably, with the top ten of large islands today. This would accord with the various description of Atlantis as a large island. By way of comparison Hatton Rockall would, if it were an island, be, roughly, treble the side of Ireland, [356] double the size of Iceland [357] and marginally smaller than Great Britain (England, Scotland and Wales) which has a surface area of approximately 230 000 km². [358]

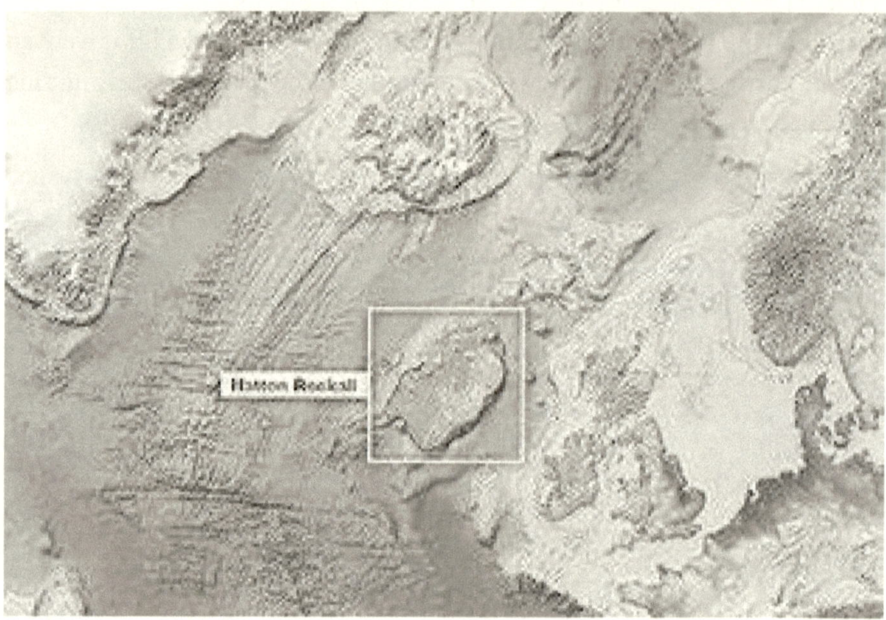

Figure 7 3 Hatton Rockall (as if an island) in context [359]

7.2.2.3   The mountains

The plain on Hatton Rockall is surrounded, on three sides, by 'mountains' which, for the most part are much steeper on the 'seaward' side than on the sides facing the plain. To the west of the plain is Hatton Bank, to the immediate north of the plain is a particularly steep extension of Hatton Bank, to the north east is the George Bligh Bank and to the east is Rockall Bank. The steep contours on the seaward side of the mountains are visible on, in particular, Figure 7.8.

Rockall Bank (which would, if Hatton Rockall were an island, form the highest point of the eastern range of mountains edging the plain) stands approximately 1 500 metres above the plain. By way of comparison the highest mountain in the United Kingdom[360] is Ben Nevis in Scotland at 1 344 metres[361] and in Ireland[362] it is Carrauntoohil at 1 038[363] metres.

In the middle of northern section of the plain, there is a small 'mountain' which now goes by the name of 'Mammal Mount'. I shall elaborate, in Section 7.3.2.4, upon the topography of Mammal Mount.

7.2.2.4   Other islands

To the north of Hatton Rockall (north of Hatton Bank) the margins of Hatton Rockall are ill-defined and merge with the foot-slopes of a sea-mount, Lousy Bank. There are, as is clear from Figure 7.3 and Figure 7.4, in addition to Lousy Bank, various other sea mounts and 'banks' to the north east of Hatton Rockall. These could, with a raised base or a lower sea level, easily have been islands.

Figure 7 4 Close up of United Kingdom, Ireland, Hatton Rockall showing sea mounts to the north east of Hatton Rockall.) [364]

This possibility is corroborated by a map provided by the National Oceanic and Atmospheric Administration of the United States

("NOAA").[365] The map shows a model of global elevations at about 18 000 years ago when mean sea level was approximately 110 metres below the present level.

The map shows (although it does not cover the exact time period referred to by Plato) that, in the recent geologic past,[366] part of Hatton Rockall was a subaerial island, that there were a number of islands in the vicinity of Hatton Rockall, that the United Kingdom and Ireland were part of the continental mainland and that the distance between the Hatton Rockall and the, then, edge of the continental mainland was much less than today. Figures 7.5 and 7.6 are extracts from that map.[367]

Could these have been the islands referred to by Plato when he said Atlantis *"was the passage to other islands"* from which one *"might pass to the whole of the opposite continent"*?[368]

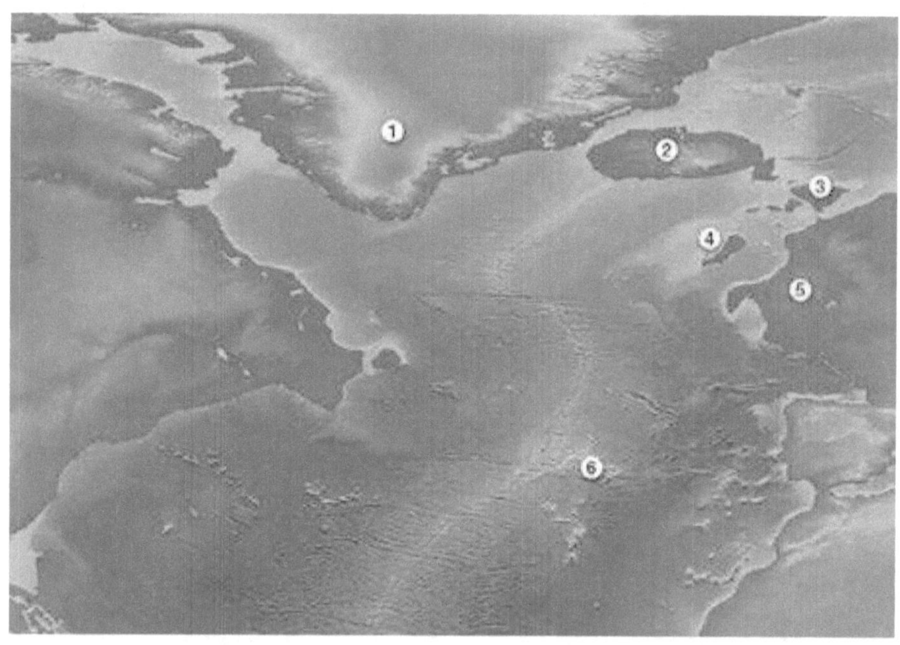

Figure 7 5 The North Atlantic at about 18 000 BP [369]

<u>*Key to Figure 7.5.*</u>
1.Greenland; 2.Iceland; 3.The Faeroe Islands; 4. Hatton Rockall;
5. Ireland; 6 The Azores

Figure 7 6 Close up map of the Hatton Rockall region showing Hatton Rockall as, in part, an island and other islands in its vicinity (At 16 000 BCE) [370]

7.3 Comparison of the topographies of the plains of Atlantis and Hatton Rockall

7.3.1 Reiteration

In the Jowett translation of the *Critias*, the plain is described as follows:

"It was for the most part rectangular and oblong, and where falling out of the straight line followed the circular ditch. The depth, and width, and length of this ditch were incredible, and gave the impression

that a work of such extent, in addition to so many others, could never have been artificial. Nevertheless I must say what I was told. It was excavated to the depth of a hundred feet, and its breadth was a stadium everywhere; it was carried round the whole of the plain, and was ten thousand stadia in length.[371] [372] and that *"Further inland, likewise, straight canals of a hundred feet in width were cut from it through the plain, and again let off into the ditch leading to the sea: these canals were at intervals of a hundred stadia, and by them they brought down the wood from the mountains to the city, and conveyed the fruits of the earth in ships, cutting transverse passages from one canal into another, and to the city."*[373, 374]

It goes on, after describing the plain, to state that *"The entire country was divided into sixty thousand lots, each of which was a square of ten stadia …"*[375,376,377]

From this may be extracted, in addition to the parts already discussed, that:

➤ the plain was, generally, rectangular,

➤ surrounding the plain there was a 'ditch' which, although he had been told that it was man made, was so extensive that it was hard to believe that it was man made and

➤ across the plain were canals which interlinked by "transverse passages" and fed into the ditch.

7.3.2   The plain and its features

7.3.2.1   The size and orientation of the plain

Hatton Rockall Basin would, if Hatton Rockall were subaerial, form the central plain of the island. It measures, approximately 444 kilometres by 185 kilometres.[378] Whilst this is somewhat smaller[379,380] than the dimensions given by Plato it is, for the most part, oriented North/South and, like that of Atlantis' plain, it is longer than it is broad.

Furthermore, if it is accepted that Hatton Rockall did sink, it is possible that the actual dimensions of the island and plain could have changed through distortion and deformation as it sank.

7.3.2.2   Marginal channels

The plain has, according to Roberts,[381] a number of unusual features. Its floor is arched and has well developed channels around its edge. The form, width and relief of these channels vary considerably.[382]

Once again it is probably easiest to, firstly, allow a diagram to speak. Figure 7.7 is a diagram showing an east/west cross section of Hatton Rockall. (See the Appendix as well.) It can be seen from this that there is a depression on both edges of the central plain (The central rise is Mammal Mount, with which I shall deal later.) and, as can also be seen simply from the perspective, the depression, or channel, is deep.

The existence of these channels is even more apparent from Figure 7.8, which is based on a map provided by Davies and Laughton.[383] That map shows that there is a channel around virtually the entire edge of the entire plain. M Sayago-Gil et al, in their article[384] on the geomorphology of parts of Hatton Rockall discuss, amongst other things, measurements of the 'contourite channels'. These contourite channels (a general term including moats, furrows and scours) vary in length from about 14 kilometres to about 38 kilometres and the moats have an average width of 1.3 kilometres.

Hatton Rockall's 'plain' therefore has a "ditch" of significant size running around most, if not all, of it.

This is, again, consistent with Plato's description of Atlantis.

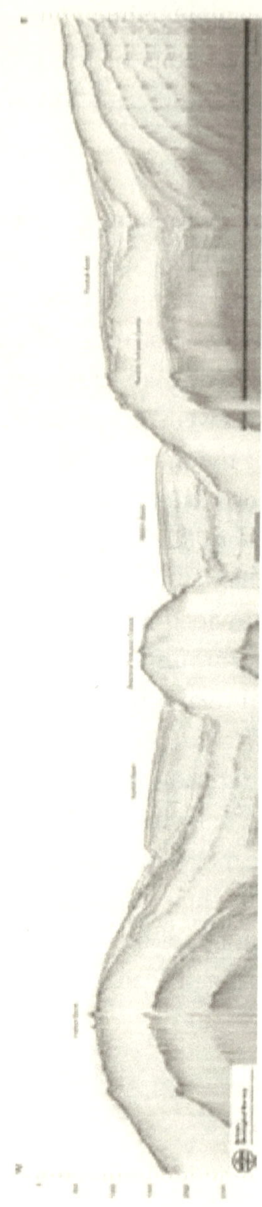

Figure 7 7 East/West cross section (rotated 900) through Hatton Rockall and the northern section of the plain. [385]

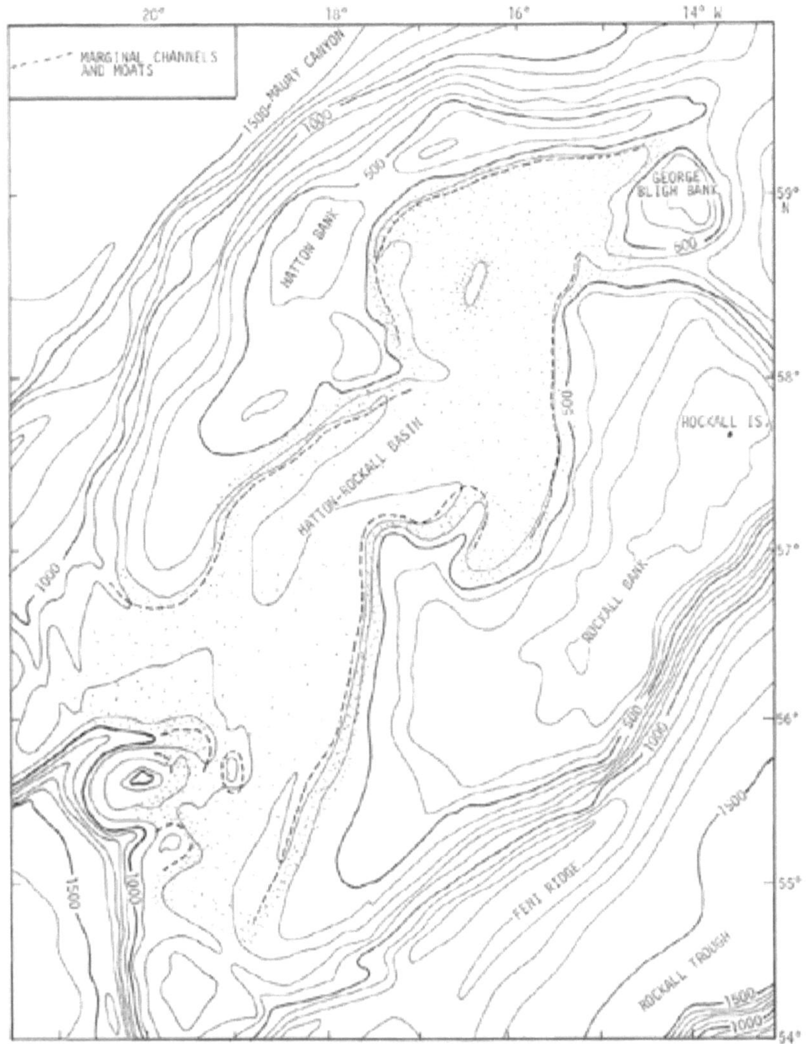

Figure 7 8 Bathymetric chart of Hatton-Rockall showing marginal channels and moats. [386]

7.3.2.3   Straight canals in the plain

The "*straight canals of a hundred feet in width were cut … through the plain*". These were "*at intervals of a hundred stadia*" and had "*transverse passages from one canal into another…* ".

The surface of Hatton Rockall's plain is divided up into natural polygons which can measure between a few hundred metres to kilometres across. These polygons, which are only found in the central part of the plain, are defined (or 'encircled') by channels (lineated depressions) of between about 5 to 10 metres deep which connect to each other in an interlinking network. These channels may be up to hundreds of feet[387] in width[388] but within the polygons the surface is largely flat.[389]

The author[390] of the blog "Thinking Deep Blue",[391] writes, when describing the polygonal faults, "*Think of the Giants Causeway of hexagonal rocks on a scale of 1 km per hexagon. The multibeam map looks like giraffe skin with these kilometre wide hexagons displaying acres of fine sand endlessly burrowed and turned over by unseen creatures. The gaps between hexagons are 30 m deeper and the sub bottom profiler shows that there are deep fault lines fracturing the earth's crust here. ... There were some little 'pimple' features on the multibeam maps that showed 3 m deep, 10 m wide depressions ... "*[392, 393]

Such natural polygons can, and do, still appear on various land surfaces today. They are, usually, found in areas which are subject to freezing and may divide up the land with channels filled with water. Photographs taken of the Kaminak Lake Area, Kivalliq, Nunavut, Canada showing frost polygons[394] can be found at http://www.prairie.illinois.edu/shilts/gallery/shilts-tyrrell.shtml and, in particular, at http://www.prairie.illinois.edu/shilts/gallery/shilts-0068.shtml and http://www.prairie.illinois.edu/shilts/gallery/shilts-0072.shtml. The area in the photographs was covered by the post glacial[395] Tyrrell Sea which was, in the area in the photographs, about 100 metres deep.[396]

It appears to me to be probable that if there are naturally occurring depressions, or gullies, criss-crossing the plain and if there were people living on Hatton-Rockall they could, and probably would, have used those gullies as described by Plato.

This being so, the existence of the polygonal faulting is another consistency with, and corroboration of, Plato's description.

7.3.2.4   The topography of Mammal Mount and its immediate surroundings.

Mammal Mount[397] is located centrally in the northern part of the plain.[398] Figure 7.9 shows its position on the plain and in relation to the whole area.

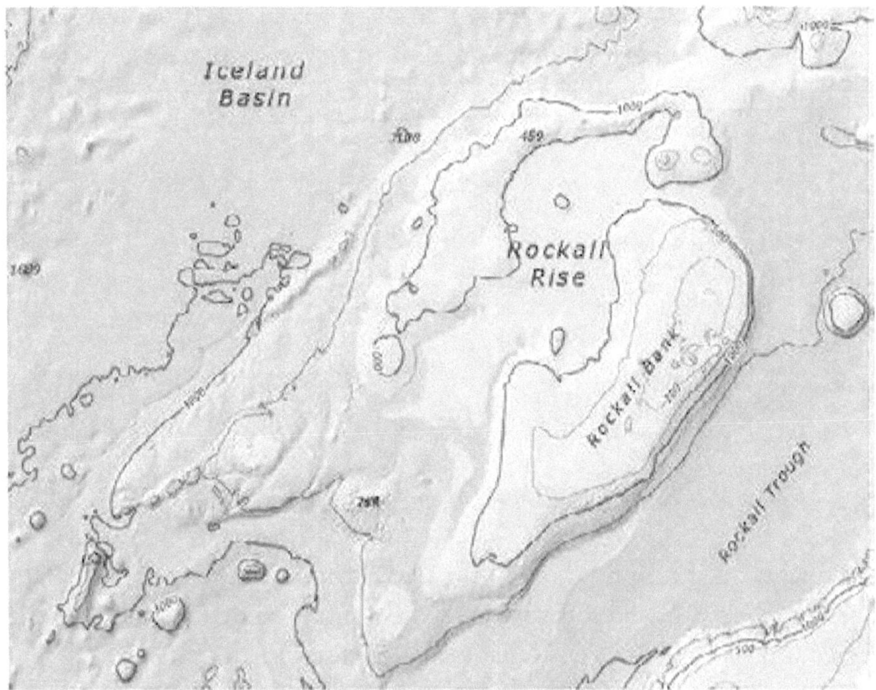

Figure 7 9 Hatton Rockall with Mammal Mount being the circular high point above the 'k' in Rockall Rise [399]

Figure 7.10 shows a cross section through the plain and Mammal Mount, also showing its central position.

The mount is a natural rock dome in the sea floor. The actual volcanic dome is about 25 kilometres in diameter and stands about 200 metres above the plain. Because of sediments covering the dome, it is probable that the actual height above the plain is higher than this.[400] At its highest point there is a pinnacle of volcanic rock which contributed to its naming. (See Figure 7.10) Surrounding Mammal Mount is a natural moat which, from seismic data, appears to be up to 100 metres deep.[401]

People of most cultures have often tended to venerate certain areas or to make use of pre-existing physical features to further and simplify their plans. Mammal Mount having the shape of a breast, or mammary gland, would have probably been eminently suitable for veneration and if anyone wanted to surround a sub-aerial Mammal Mount with a water filled moat it would have been logical to use the pre-existing natural moat.

Similarly, it is equally possible that, if the moat was a natural feature and a prehistoric person wished to explain how a mountain came to be surrounded by a water filled moat, he, or she, would ascribe it to the actions of a god.

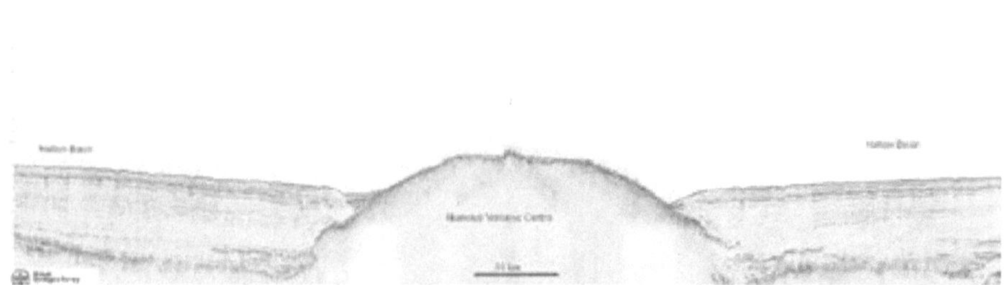

Figure 7 10 Detail of Mammal igneous complex[402]

In conclusion therefore, there is, the middle of the Hatton Rockall plain, a small mountain which is surrounded by a natural depression, or moat. Again, this is consistent with Plato's description of Atlantis.

7.4   Conclusion

It is therefore apparent that Hatton Rockall would, if it were an island, be a large island with a large central plain, the plain is rectangular and primarily north/south orientated, has a large channel running around its edge, has many smaller channels criss-crossing its surface, has a central small mountain with a natural moat and is surrounded, on three sides, by high mountains which dip precipitously on the seaward side.

This is all entirely consistent with Plato's description of, the now sunken, Atlantis.

SECTION EIGHT: HATTON ROCKALL AS AN ISLAND AND ITS POSSIBLE CLIMATE

8.1   Introduction

If Atlantis was what is now Hatton Rockall it follows that it must, within orally recorded history, have been above sea level; an island. This is not something which is presently accepted by geologists, or the scientific community in general, however, given that the similarities go beyond mere coincidence, this must be considered further.

I shall, in this section, consider,

> firstly, certain facts which may indicate that Hatton Rockall could have been sub-aerial at the relevant time and

> secondly, I shall consider the possible palaeoclimate of Hatton Rockall as an island and, more specifically, whether it could have had, or did have, a temperate climate.

8.2   A subaerial Hatton Rockall

8.2.1   Is there anything to substantiate that Hatton Rockall was subaerial during the late Pleistocene and early Holocene?

I have already mentioned, in Section 7.2.2.4, that it is accepted that at the least a portion of Hatton Rockall was an island at around 16 000 BCE. I shall now consider other factors which indicate the possibility that deeper portions of it were subaerial.

8.2.1.1   Carbonate mounds

In the North West Atlantic, in particular, off the Irish Shelf and on and around Hatton Rockall, there are large numbers of 'carbonate mounds' - generally grouped into so-called provinces.[403] These carbonate mounds can extend up to hundreds of metres high and a few kilometres across.[404] They have been formed by the deposition of calcium carbonate by colonies of cold water corals; corals which can only grow and thrive in deep water. These mounds are, it is important to note, presently found at depths ranging from 500 metres to 1000 metres.

Studies have been made of the mounds and cores have been taken in order to, amongst other things, ascertain their age.[405,406] The cores cover the time period from the present, backwards in time, to the early Pleistocene or even earlier – a timespan of roughly 2 500 000 years. What is significant, for our purposes, about these cores is that it is apparent that there are substantial hiatuses,[407] or breaks, in the development of the mounds. These hiatuses, the authors say, reflect periods of non-deposition of the calcium carbonate[408,409,410] due, presumably, to the death of the corals. They span a total of up to 200 000 years. The depths at which these mounds are found are not depths which are (as was seen in section 7.2.2.4) accepted as having been above sea level in the late Pleistocene. Their deaths cannot, as such, be attributed merely to the lower sea level cause by the water being 'locked up' in ice caps.

One possible cause for the death of the corals (and consequent non-deposition of the calcium carbonate) would be if Hatton Rockall, together with the mounds, had been lifted out of the water. It is therefore particularly significant, firstly, that the deaths of the corals causing these hiatuses have all been during glacial periods[411,412] and, secondly, that this is the reverse of what has been found with similar corals in other parts of the world (i.e. the growth of the corals in other areas has predominated in glacial periods[413,414]).

This death of the corals in glacial periods accords with my proposal (which I discuss more fully in Section 13) that Hatton Rockall, being on the periphery of the ice sheets, was lifted above sea level by the weight of the ice sheets and subsequently subsided, largely by isostatic adjustment, after the ice sheets melted.

Insofar as it might be questioned whether similarly formed carbonate mounds can be, or are, found on land today the answer is yes. One example of this is the so-called Kess Kess mounds in the Hamar Laghdad area, Anti-Atlas, S.E. Morocco.[415,416]

8.2.1.2   Gypsum

A further possible corroboration for the proposition that the Hatton Rockall area was above water may be found in the fact that gypsum has been found in at least one geographically close mound - in Mound Perseverance.

The significance of this finding is that gypsum is normally formed above water – it is a classic evaporitic mineral. Pirlet *et al*,[417] who studied this, acknowledge that the finding of gypsum in this situation is intriguing but go on to propose a process by which it could have been formed underwater.

Whilst I cannot deny that it is possible for the gypsum to have been formed underwater (that is beyond my knowledge) I propose that, using 'Ockham's razor', it is more logical that it was formed above water and has later submerged. This would also be consistent with my proposal.

8.2.1.3   Sediment

Sediment is frequently laid down by water and its nature can provide a clue to the history of the site. Changes or breaks in the sediment, referred to as unconformities, can also provide proof of either changes in the area of deposition or the source of the sediment. If, for example, land that was beneath water was (for whatever reason) no longer covered one would expect to find a change, or unconformity, in the layers of sediment laid down and a reduction in the amount of sediment for the period that it was dry land.

What is the situation at Hatton Rockall?

There are definite unconformities and suggestions that Hatton Rockall might, over extended time periods, have been uplifted and have been subaerial. This appears to be relatively uncontentious. Where it becomes an unknown is with regard to whether or not Hatton Rockall was subaerial during the Pleistocene and/or early Holocene. If, as is suggested by the carbonate mounds and gypsum, Hatton Rockall was subaerial during, at least, the Pleistocene one could reasonably expect that there was reduced sedimentation.

In fact, the most recent sedimentary layers in the Hatton Rockall basin are thin. In the article *The geology of the UK Hatton-Rockall margin* the author refers to the sediment laid down during the Plio-Pleistocene epochs (i.e. the last, approximately, 5 million years) as being a 'veneer'. Similarly Stoker *et al*[418] refer to a 'thin layer' of sediment having been laid down during the last, approximately, 2.5 million years (the Quaternary Period).

Finding such thin layers of sediment is therefore consistent with Hatton Rockall having been above the surface of the sea during the Pleistocene.

8.2.2   Is Hatton Rockall still sinking?

Although the basis for such a suggestion is tenuous there is, possibly, a basis for it. (This is also discussed, from another angle, in sections 13.4 and 10.5.5.)

According to measurements performed by Nick Hancock during 2014 the height, above mean sea level, of the of the top of Rockall Islet has, compared to the results of the calculations done in 1977, decreased by about 85cm.[419] The other measurements from north to south and from east to west have also decreased. The position of the islet (as determined in 1977) has, however, not changed as a result of the new measurements.

This would be consistent with subsidence, due to isostatic adjustment, also found in western Ireland.

8.2.3   Conclusion

Although the information is scant[420] it does appear to be possible that Hatton Rockall might have been above the sea, and an island, at the end of the Pleistocene / beginning of the Holocene and, also, that it might still be sinking.[421]

8.3   The Climate

Is it possible that Hatton Rockall, if it had been an island during the late Pleistocene, could have had a climate similar to that described by Plato? We will now consider this issue.

8.3.1   Climate – terminology and concepts

Before dealing with the specifics I will set out certain general concepts relating to climate, starting with a definition of climate itself.

Climate, as defined by Wikipedia, is *"the long-term pattern of weather in a particular area."* Climate is different from weather *"in that weather only describes the short-term conditions of these variables in a given region."* *"The difference between climate and weather is usefully summarized by the popular phrase 'Climate is what you expect, weather is what you get.'"*[422]

There are numerous factors which contribute to determining the climate of a particular geographical area – this can even extend down to the very local level. *"The climate system is a complex, interactive system consisting of the atmosphere, land surface, snow and ice, oceans and other bodies of water, and living things."*[423]

The main factors affecting the climate of an area, albeit not necessarily in this order, are: <u>L</u>atitude, <u>O</u>cean currents, <u>W</u>ind and air masses, <u>E</u>levation, surrounding <u>R</u>elief (or topography) and proximity, or <u>N</u>earness, to large bodies of water.[424,425] This is sometimes referred to by the acronym 'LOWERN'. The climate is also be affected by the Earth's tilt and its natural movement, over millennia, nearer to or further away from the sun.[426]

Climates can, and have in the past, changed dramatically and rapidly[427] – in even as little as ten years.

8.3.2   Reiteration and introduction

Having briefly considered the general climate of Atlantis in Section 4, I shall now do a more detailed comparison.

To reiterate, Atlantis

> ➢ received most of its rainfall in winter

> ➢ had lush natural (and cultivated) growth and vegetation with a temperate, or warm, climate and

> ➢ benefited from the protection afforded by the mountains from the "northern blasts".

I shall first consider the climate of Hatton Rockall today, then certain scientific authorities and articles which reconstruct the climate at around the time that Atlantis is reported to have sunk and, thereafter, the climates prevailing in Ireland and parts of the United Kingdom today.

Ireland and the United Kingdom, are the closest significant land masses to the Hatton Rockall area and are both, like Hatton Rockall, particularly if it were an island, affected by the warm Gulf Stream ocean current, a warm surface current originating in the Gulf of Mexico which flows north-east across the Atlantic. (Note that when I refer to the 'Gulf Stream' I do not do so in a strict, or oceanographic, sense but include within that concept all the various warm currents which branch off from it or into which it changes.)

I shall also, simply for comparison's sake, briefly consider the climate of the Faeroe Islands.

8.3.3   *The Hatton Rockall area today*

The Hatton Rockall area today is known for its inclement weather and has a fearsome reputation amongst, particularly, fishermen and the oceanographic community.[428]

The area receives plentiful rain, with most of it falling during the winter months.[429] This is consistent with the description of Atlantis. The climate of the region is strongly influenced by a large-scale westerly air circulation and by the Gulf Stream (See Figures 8.1 and 8.2.) which brings humid, mild air to the region and is largely responsible for the temperate climate of Ireland, southern England and north-west Europe.[430]

Today's climate of Hatton Rockall does not resemble the temperate climate described by Plato, but one must bear in mind that, if it was not submerged, its mere existence as an island would have changed the local climate. Hatton Rockall as an island would, as is apparent from Figure 8.2, be 'embraced' on virtually all sides by warm currents. This would have a significant warming and moderating effect on its climate and if Hatton Rockall were an island today its climate would be substantially more equable than it is at present and probably something similar to that of Ireland and also of the western portions of England.

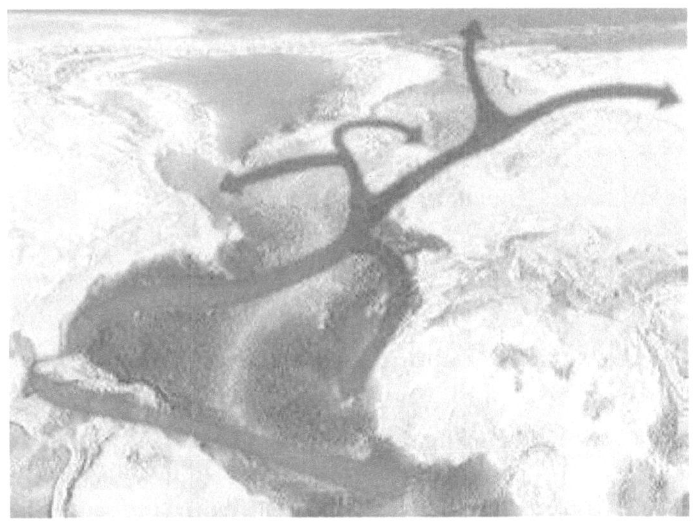

Figure 8.1 The Gulf Stream and related currents, at present, in broad outline[431]

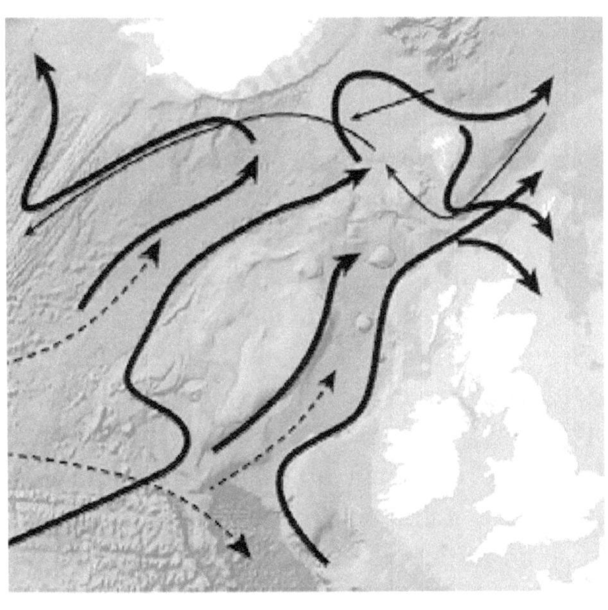

Figure 8.2 Detailed map of present currents in the N.E. Atlantic[432]

8.3.4   *The Palaeoclimate of the North Atlantic region*

What then, accepting that the present climate of the area is not a good basis upon which to judge the climate of the area in the past, was the climate of the region at the beginning of the Holocene?

Apart from any general warming prior to the start of the Holocene, Li and Battisti[433] found, in their simulation of the climate during the last glacial maximum, that at that time (i.e. when the sea ice cover should have been at its maximum) sea ice did not extend south of 60^0 N on the eastern edge of the Atlantic. Their findings were also supported by other proxy data. The implication of this is that, that even at the time of maximum ice, Hatton Rockall would, if it were an island, have been ice free. (The northern tip of Hatton Rockall is, I reiterate, south of 60^0 N.) The possibility that the sea ice did not extend as far south as Hatton Rockall at the relevant time may also be contemplated in the light of the findings of Alonso-Garcia *et al.*[434] – albeit that the time period covered by their study is much earlier. Li and Battisti also found that the sea surface temperatures in the area which includes Hatton Rockall were, compared with earlier simulations, 1 to 3^0 C warmer than had previously been accepted and, also, that the glacial climate probably had, contrary to expectations, weaker storms and was "quiescent" compared to today's climate.

Zhirov[435] suggests that it is possible that an Atlantis-like land mass (which he does, however, not place in the position of Hatton Rockall) could have changed the flow of the Gulf Stream and its extensions and thereby changed and warmed the climate in certain areas. The possibility that the Gulf Stream, and/or its offshoots, might have been attenuated or redirected also appears to be contemplated by the findings of Ezat *et al.*[436] and Thornalley *et al.*[437] – although, in these cases, they propose that it was caused by developments in the Arctic.

The last glacial maximum[438] was about 10 000 years prior to the destruction of Atlantis and it was after this that the world's climate started to warm. The improvement of the climate was rapid, particularly between about 16 000 to 15 000 years ago and throughout Western Europe and the North European Plain. Human occupation recommenced in northern France, Belgium, north-western Germany, and southern Britain between 15 500 to 14 000 years ago and by between 12 000 to 10 000 years ago anatomically modern humans began to settle on the western coast of Norway and southern Sweden up to latitude 65° north.[439] At about 12 700 BCE there was a further warming trend (the Bølling-Allerød interstadial period) in, at least, the northern hemisphere. It is noteworthy that the data for the time, and specifically in western Ireland, suggest that the period there was as warm as today.[440,441]

About 2 000 years after the warming commenced (which, in geological time frames, is short) there was in the northern hemisphere a sudden drop in temperatures. This was the start of a period of cold climatic conditions which lasted approximately 1 000 years between approximately 10 800 and 9 500 years BCE.[442,443] This was known as the 'Younger Dryas'.[444] (There is no certainty as to what caused the Younger Dryas but is has recently been strongly advanced[445] that it was due to a bolide or meteorite strike.)

Rasmussen *et al.*[446], in a 2016 study, found that there was a 'warm blob' in the North Atlantic which extended north to Greenland and Iceland – up to the edge of the ice pack. In the area of their study, which probably included the Hatton Rockall area, the climate changes were gradual[447] and the climate of an island within this area would presumably, because of the warm sea, have been more equable.

Furthermore, according to a study[448] by Hu *et al.*, there would (prior to the opening of the Bering Strait in about 10 000 BCE) have been, an annual average warming of up to 1.5^0 Celsius in the north Atlantic, the Arctic and other surrounding areas. This would have been due to reduced circulation and a strengthening of the meridional overturning circulation in the North Atlantic Ocean.

Finally, although it was a study covering the opposite side of the Arctic sea, according to Elias *et al,*[449] late glacial (i.e. between about 12 000 and 9 000 BCE[450]) insect assemblages from the Beringia region reflect a climatic warming from glacial conditions to summers that were warmer than at present.

Whilst I have quoted dates for the beginning and ending of events it must be understood that the timing of events in prehistory can be difficult to ascertain with precision. Even using different ice cores from Greenland, which one might think would be easy to calibrate, have resulted in offsets of up to several thousand years in the glacial periods and even noteworthy differences in the Holocene.[451]

Thus, whilst there is no error-free date[452] for the ending of the Younger Dryas period, it is generally accepted as having ended at about 9 500 BCE.[453,454] The end of the Younger Dryas (which would, therefore, have occurred around the time given by Plato for the flooding of Atlantis) came about abruptly with a warming of between 5 to 10 $^{\circ}$C. over an approximately 40 year period, less than he lifetime of an average person today.[455, 456]

In conclusion it therefore appears that the climate of the Hatton Rockall area, at the time of the destruction of Atlantis and, possibly, for a few thousand years before that, was not as inclement as it is now was, in fact, temperate and may even, if some areas in England and Ireland are anything to go by, have had a sub-tropical climate.

8.3.5   *Current climates of nearby islands*

8.3.5.1   Ireland's climate

Ireland has a mild oceanic climate and experiences few extremes[457] with, generally, cool, but not cold, winters and warm, but not hot, summers.[458] Rainfall is prevalent during winter months with the south western areas experiencing the most rainfall as a result of south westerly winds.[459]

Its climate is typically insular and temperate, avoiding the extremes in temperature of many other areas in the world at similar latitudes.[460] As a consequence of its mild climate and frequent rainfall Ireland has lush vegetation and this has earned its sobriquet the 'Emerald Isle'.

The climate of County Kerry, in South West Ireland, is such *"that subtropical plants such as the strawberry tree*[461] *and tree ferns, not normally found in Northern Europe, thrive in the area".*[462] Other *"sub-tropical species such as palm trees"*[463] also grow in Ireland as a result of its climate.

Geoffrey Keating, the author of *The History of Ireland*[464], quotes, at page 110, the semi-historical figure Ioth saying *"and he thereupon proceeded to praise Ireland, and said it was wrong for them to dispute with one another, seeing that the island so abounded in honey, in fruit, in fish, and in milk, in grain and corn, and that the climate was so temperate as regards heat and cold."*

These descriptions are very similar, in tenor, to Plato's description of Atlantis.

8.3.5.2   United Kingdom's climate

I shall concentrate on the south west of the United Kingdom, primarily Cornwall, Devon and parts of Wales because they are (like Hatton Rockall would be if it were an island) the most affected by the Gulf Stream and its various offshoots.

The climate of south west England is, similarly, described as "oceanic ... [which] *is typified by cool winters with warmer summers and precipitation all year round, with more experienced in winter. ... Summer average maxima range from 18° C (64 F) to 22° C (72° F) and winter average minima range from 1° C (34° F) to 4° C (39° F) across the south-west.* "[465]

Cornwall's climate, specifically, is described as *"a temperate oceanic climate ... and has the mildest and sunniest climate in the United Kingdom, as a result of its southerly latitude and the influence of the Gulf Stream. The average annual temperature in Cornwall ranges from 11.6 °C (53 °F) on the Isles of Scilly to 9.8 °C (52 °F) in the central uplands. Winters are amongst the warmest in the country due to the southerly latitude and moderating effects of the warm ocean currents, and frost and snow are very rare at the coast and are also rare in the central upland areas. ... The moist, mild air coming from the south west brings higher amounts of rainfall than in eastern Great Britain, at 1050 to 1290 mm (41.4 to 50.8 in) per year, however not as much as in more northern areas of the west coast. The Gulf Stream, bringing warm air from the Caribbean north-east toward Europe, makes Cornwall's weather distinctly milder than other places in the world at the same latitude, such as Newfoundland"*[466].

Cornwall has, due to the effects of the Gulf Stream, the United Kingdom's *"only area of sub-tropical climate at the extreme south-west of Cornwall and the Isles of Scilly"*[467] where palm trees are a common sight. This has also resulted in a number of well-known botanical gardens, such as Trebah and the Lost Gardens of Heligan, in this area.

Devon, similarly, *"has a mild climate, heavily influenced by the North Atlantic Drift. ... The county has warm summers with occasional hot spells and cool rainy periods. Winters are generally mild and the county often experiences some of the mildest winters in the world for its latitude ..."*.[468]

Wales, whilst lying further north than Cornwall, Devon or Somerset also has a mild climate, influenced by the warm North Atlantic Drift extension of the Gulf Stream. It "*lies within the north temperate zone. … Welsh weather is often cloudy, wet and windy, with warm summers and mild winters.*" and "*Temperatures in Wales are kept higher than would otherwise be expected at its latitude by the North Atlantic Drift, a branch of the Gulf Stream. The ocean current, bringing warmer water to northerly latitudes, has a similar effect on most of north western Europe. As well as its influence on Wales' coastal areas, air warmed by the Gulf Stream is carried further inland by the prevailing winds.*"[469]

All of these areas, along with the rest of the United Kingdom, receive most of their rain in the winter months.

8.3.5.3   The Faeroe Islands' climate

The Faeroe Islands, which lie to the north of Hatton Rockall (at approximately 62⁰ N) have a climate which, although also affected by the Gulf Stream, is very different. Its "*climate is classed as Maritime Subarctic… . The overall character of the islands' climate is influenced by the strong warming influence of the Atlantic Ocean, which produces the North Atlantic Current. … The islands are windy, cloudy and cool throughout the year with over 260 annual rainy days. The islands lie in the path of depressions moving north-east and this means that strong winds and heavy rain are possible at all times of the year. Sunny days are rare and overcast days are common.*"[470]

8.3.6   Conclusion

There is, therefore, a reasonable possibility that the climate of Hatton Rockall, as an island, could have matched (at *circa* 9 500 BCE and before) the climate described as having existed at Atlantis.

SECTION NINE:
WHAT, OR WHERE, WAS GADES / GADEIRA?

9.1   Reiteration and introduction

Plato stated, in the *Critias*,[471] that Poseidon gave to his first born son, Atlas (after whom the island and ocean were named), " ... *his mother's dwelling and the surrounding allotment ... .*" and "*To his twin brother, ... the extremity of the island towards the Pillars of Heracles, facing the country which is now called the region of Gades*[472],[473] *in that part of the world ....*" The precise meaning of the words *"the extremity of the island ... facing the country which is now called the region of Gades"* is not entirely clear. The Greek word translated as "Gades" by Jowett is also translated, in the Perseus *Critias*, as "Gadeira".[474] The impression given, in both in the Jowett and Perseus Tufts translations, is that a portion of the island of Atlantis faced Gades/Gadeira and that it, Gades/Gadeira, and Atlantis were in relatively close proximity.

I shall, for the sake of simplicity, from here onwards (unless quoting from authors who use the word Gades) refer to Gades/Gadeira simply as Gadeira.

Before going further it should be noted that, today, a reference to Gades is generally, and fairly consistently, accepted as a reference to modern Cadiz in Spain. It is for this reason that many people seeking Atlantis have sought it near modern day Cadiz. Cadiz is, however, not anywhere near Hatton Rockall and so, if Hatton Rockall was Atlantis then the reference to Gadeira cannot be a reference to modern Cadiz. There must be another logical explanation.

Place names do change over time and this fact is recognised, implicitly, by Plato. He stated that the region *"is now called"* Gadeira. It follows, by necessary implication, that Gadeira was not always known by that name. It should also be borne in mind that the description of Atlantis has gone through a number of languages – at least 3 and easily 4[475] and also that it is possible that there were two or more Gades / Gadeiras at that time, or over a period of, time. (Just as there are multiple Cadiz's,[476] Londons, Paris', Athens' and other towns and cities today.)

9.2   Could Ireland be Gadeira?

Accepting, at least for the moment, that Hatton Rockall was Atlantis then the other land (in relatively close proximity) towards which one could face, from the southern extremity, would be present day Ireland and the continental shelf surrounding it. (See map, Figure 9.1.) Bear in mind that, at the ostensible time of Atlantis, Ireland was not an island but formed part of what was then still the European mainland. (The continental shelf surrounding Ireland and the United Kingdom was sub-aerial at the time.)

The questions then raised are

➢ is there any possibility that 'Gadeira' might have been what is now known as Ireland and, more particularly,

➢ was Gadeira, if not at the time of Atlantis then by the time considered by Plato, an island?

In pursuing this line of thought I shall refer to certain works of ancient historians and other classical literature as well as to Irish mythology.

In considering the possible Ireland/Gadeira link I shall also briefly consider to the link between Gadeira and Erytheia. Gadeira has, over time, been associated with the mythical Erytheia which has, in turn, also been linked with the Fortunate Isles[477] or the Islands of the Blest.

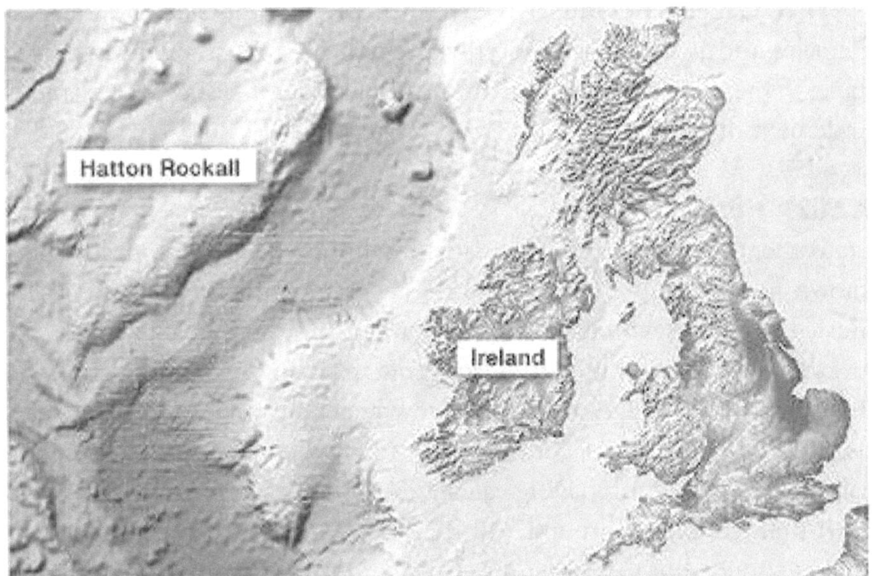

Figure 9.1 The United Kingdom, Ireland and a portion of Hatton Rockall [478]

9.2.1 Ancient Greek writers

I shall refer to the various ancient writers in chronological order.

9.2.1.1 Herodotus

Herodotus was an ancient Greek historian who lived in the 5th century BCE (c. 484 BCE – c. 425 BCE). He, therefore, preceded Plato in time and died around the time that Plato is believed to have been born. Herodotus is sometimes known as the "Father of History" as he was the first historian who is known to have collected his materials systematically, to have tested the accuracy of the collected material and to have arranged them into a narrative.[479]

In Book 4, Chapter 8 of his work, the *Histories*[480], he (when discussing the labours of Hercules) wrote: "*Heracles, driving the cattle of Geryones, came to this land, which was then desolate, but is now inhabited by the Scythians.*[481] *Geryones lived west of the Pontus, settled in the island called by the Greeks Erythea, on the shore of Ocean near Gadira, outside the Pillars of Heracles.*"

Here Gadeira ("Gadira") is described as being outside the Pillars of Heracles and is linked with Erytheia. Erytheia was accepted as being in the extreme west. It is not clear from this whether 'Gadira' is an island,[482] settlement or region.

9.2.1.2   Eratosthenes

Eratosthenes was a Greek polymath, the librarian in Alexandria, who is known for his study of mathematics, geography, poetry, astronomy and music theory and who lived from about 276 BCE to about 195 BCE. He was the first person to use the word "geography" in Greek and he invented the discipline of geography as it is now generally understood. He made a number of important contributions to mathematics and science and has been credited with having calculated, to a fair degree of accuracy, the Earth's circumference around 240 BC.[483]

He wrote a number of works of which one, *Geographica*, is important here. This work has, unfortunately, been lost but fragments are retained in the works of other authors making reference to it. Strabo[484] (who is also mentioned below) is one of these authors.

Strabo, albeit disagreeing[485] with Eratosthenes, states[486] that, according to Eratosthenes, Gadeira is five days sailing distant from the Sacred Promontory.

It is uncertain as to which sacred promontory he was referring or where the sacred promontory was. According to Wikipedia, the terms "*Sacred promontory or sacred cape or holy promontory or cape was a name assigned by the ancient Greeks and Romans to salient promontories extending into large bodies of water at strategic locations, typically containing a temple to the god of the sea.*"[487] Of the sacred promontories listed by Wikipedia there are only two in the Atlantic which can be considered here,[488] namely Cape Roca and Cape St Vincent. Cape Roca is near Lisbon, Portugal and Cape St Vincent is to the south of Lisbon.

It is impossible to know the route taken by the sailors referred to by Eratosthenes or the speed at which they sailed however, according to the website www.sea-distances.org, a vessel sailing from Lisbon, Portugal to Cork in Ireland would take

➢ just short of 7 days at a speed of 5 knots and

> about 5 days 17 hours at 6 knots.

The distance sailed in such a journey is about 821 nautical miles.

This may be corroborated by a statement by the author of the book *Ancient Boats in North-West Europe: The archaeology of water transport to AD 1500*.[489] He states that it was possible, in ancient times in fair weather, to travel a similar distance (about 830 nautical miles) within 6 to 7 days.

If my hypothesis that Gadeira was Ireland is correct then Eratosthenes' statement that Gadeira was five days sailing distant from the Sacred Promontory might be correct.

9.2.1.3   Appollodorus

The *Library of Appollodorus* is an ancient Greek compilation of myths and legends in three books.[490]

It is uncertain when the work was compiled, but, from references appearing in it, it could not have been earlier than the first century BCE.[491] According to the Perseus Encyclopedia[492] Appollodorus' Library, notwithstanding the fact that it was written not earlier than the first century BCE, provides a view of Greek mythology as it appeared in the archaic and classical period. The Perseus Encyclopedia also states that the most important lost source of information used in the compilation of the Library was Pherecydes of Athens,[493] an author of a long work on Greek mythology who lived in the fifth century BCE. It is therefore apparent that, although the Library was written after Plato, the source of the information derived from a source pre-dating Plato.

In Book 2, Section 5.10[494] the Appollodorus states, *"As a tenth labour he [Heracles] was ordered to fetch the kine of Geryon from Erytheia. Now Erytheia was an island near the ocean; it is now called Gadira."* … and *"When he reached Erytheia he camped on Mount Atlas."*[495]

Pausanias, an ancient Greek traveller and geographer of the second century CE, also linked Geryon to Gadeira in his *Description of Greece*.[496]

From this it is clear that Gadeira was an island and that, according to Appollodorus, it was previously known as Erytheia. This is confirmed by Strabo in his work Geographica, referred to below.

9.2.1.4   Strabo

Strabo was a Greek historian, geographer and philosopher who lived from about 64/63 BCE until about 24 CE. He is most famous for his, 17 volume, work *Geographica*, which presents a history of different peoples and places known to his era.[497]

In Book 3, Chapter 2, Section 11[498] he states "*The ancients seem to have called the Baetis River[499] "Tartessus"; and to have called Gades and the adjoining islands "Erytheia", … "* and that "*Erytheia is called 'Blest Isle.'*"

He also, in Book 3, Chapter 2, section 12, states, albeit critically, that Homer considered Tartessis to be the farthest country towards the west, where, the radiant sun sank into the ocean.[500]

It is also of note that, similar to Atlantis, Erytheia in referred to in 'glowing' terms – the "*Blest Isle*".

9.2.1.5   Conclusion

Given the above it seems reasonable to accept that there was at least one Gadeira which was an island (possibly also known as Erytheia) and that it was in the far west of the known Atlantic ocean.

9.2.2   *Irish mythology and history*

All presently existing written Irish histories and mythologies were written down in the Christian era, often by monks, and are, as such, 'tainted' by a desire to comply with Biblical teachings. That said the histories do purport to cover ancient times even going back to events which are said to have preceded the flood of Noah.

Most myths are, by their very nature and age, uncertain and there are, often, different versions which do not agree with each other. This is so for Irish mythology as well; but one common thread running through the myths is that there have, over the millennia, been a number of invasions and settlements of Ireland.

There are many works devoted to the history, and invasions, of Ireland but one of the most succinct summaries of early Irish history which I have come across is in a book by the name of "*A Popular History of Ireland: from the Earliest Period to the Emancipation of the Catholics*"[501] by Thomas D'Arcy McGee. In the first chapter, entitled "The First Inhabitants", he wrote:

"Who were the first inhabitants of this Island, it is impossible to say, but we know it was inhabited at a very early period of the world's lifetime--probably as early as the time when Solomon the Wise, sat in Jerusalem on the throne of his father David. As we should not altogether reject, though neither are we bound to believe, the wild and uncertain traditions of which we have neither documentary nor monumental evidence, we will glance over rapidly what the old Bards and Story-tellers have handed down to us concerning Ireland before it became Christian.

The first story they tell is, that about three hundred years after the Universal Deluge, Partholan, of the stock of Japhet, sailed down the Mediterranean, 'leaving Spain on the right hand,' and holding bravely on his course, reached the shores of the wooded western Island. This Partholan, they tell us, was a double parricide, having killed his father and mother before leaving his native country, for which horrible crimes, as the Bards very morally conclude, his posterity were fated never to possess the land. After a long interval, and when they were greatly increased in numbers, they were cut off to the last man, by a dreadful pestilence.

The story of the second immigration is almost as vague as that of the first. The leader this time is called Nemedh, and his route is described as leading from the shores of the Black Sea, across what is now Russia in Europe, to the Baltic Sea, and from the Baltic to Ireland. He is said to have built two royal forts, and to have "cleared twelve plains of wood" while in Ireland. He and his posterity were constantly at war, with a terrible race of Formorians, or Sea Kings, descendants of Ham, who had fled from northern Africa to the western islands for refuge from their enemies, the sons of Shem. At length the Formorians prevailed, and the children of the second immigration were either slain or driven into exile, from which some of their posterity returned long afterwards, and again disputed the country, under two different denominations.

The Firbolgs or Belgae are the third immigration. They were victorious under their chiefs, the five sons of Dela, and divided the island into five portions. But they lived in days when the earth--the known parts of it at least--was being eagerly scrambled for by the overflowing hosts of Asia, and they were not long left in undisputed possession of so tempting a prize. Another expedition, claiming descent from the common ancestor, Nemedh, arrived to contest their supremacy. These last--the fourth immigration--are depicted to us as accomplished soothsayers and necromancers who came out of Greece. They could quell storms; cure diseases; work in metals; foretell future events; forge magical weapons; and raise the dead to life; they are called the Tuatha de Danans, and by their supernatural power, as well as by virtue of "the Lia Fail," or fabled "stone of destiny," they subdued their Belgic kinsmen, and exercised sovereignty over them, till they in turn were displaced by the Gaelic, or fifth immigration.

This fifth and final colony called themselves alternately, or at different periods of their history, Gael, from one of their remote ancestors; Milesians, from the immediate projector of their emigration; or Scoti, from Scota, the mother of Milesius. They came from Spain under the leadership of the sons of Milesius, whom they had lost during their temporary sojourn in that country. In vain the skilful Tuatha surrounded themselves and their coveted island with magic-made tempest and terrors; in vain they reduced it in size so as to be almost invisible from sea; Amergin, one of the sons of Milesius, was a Druid skilled in all the arts of the east, and led by his wise counsels, his brothers countermined the magicians, and beat them at their own weapons."

9.2.2.1   The Milesians – Gathelus and Gadelians

It is the fifth invading force in which I am particularly interested.

In assessing the relevance of this fifth 'colonisation' to the question of the names of Gadeira we will have to consider, and then bear in mind, etymologies of names.[502,503]

This last invading force is commonly known, as is indicated by McGee, as the Milesians[504]. There are suggested etymologies of the name but there is no certainty. Some say that the name derives from the Latin words for soldiers of Spain,[505] others refer to the Milesians as being the sons of Milesius.[506] What is, however, important in my view is the statement that they also "*called themselves … Gael, from one of their remote ancestors … .*"

The Gaels are, according to mythology, the descendants of Goídel Glas[507] whose name is Latinised as "Gathelus",[508] Gaidelus[509] or Gadelus.[510] Today the Irish, Scottish and similar languages are referred to as "Goidelic languages" however this was not always so. In earlier centuries they were referred to as Gadelic languages and Gadelic peoples. Hely, in his translation of O'Flaherty's work *Ogygia or a Chronological Account of Irish events,*[511] comments that "*The vowels, indeed, are very often used promiscuously and indifferently according to the diversity of dialects;*".[512] To, almost, prove this comment the statement is made,[513] in reference to Goídel Glas and his descendants, that "*… and by her had Gathel, or with more propriety, Gaidel; from whom our ancestors are denominated the Gaidelian race, and their posterity, Gaidelians and Scots.*"

From this we can draw that, at least according to Irish mythology,

> ➤ the last invaders of Ireland were descendants of a man known, in the Latinised form of his name, as Gathelus or Gaidelus

> ➤ the language spoken by these last invaders and their name were in earlier times, respectively, Gadelic and Gadelians.

Accepting that the Irish were, in earlier times,[514] known by a name with a distinct similarity to the names used by Plato does not, of itself, mean that the island was also known by the same, or a similar name. This begs the question "Is there any reference to Ireland being named after Gathelus?

9.2.2.2   Was Ireland ever named after Gathelus?

In the 1587 edition of *Holinshed's Chronicles of England, Scotland, and Ireland*[515,516] *the authors wrote about how Gathelus, the son of one Neale (who was a great lord in "Grecia"), was exiled from his country. He moved to Egypt, from there to Portugal*[517] *and then to Galicia. Ultimately, after the population of Gathelus' group had grown so much that Galicia was not able to sustain them, he and members of his group, moved to Ireland.*

As an aside (and to the extent that such extensive travelling might appear improbable so early in history) recent studies have shown that by, about, 5400 BCE Neolithic peoples sailed around the Mediterranean and had travelled as far as the western coast of the Iberian Peninsula (i.e. Portugal).[518] The Milesians arrived in Ireland, as will be seen in the next sub-section, no earlier than 2000 BCE.

After arriving in Ireland Gathelus, amongst other things, enriched the Irish language and taught the inhabitants "*letters*" and, as a result, "*... he was so acceptable to them, that to gratifie such a benefactor, they agreed to name the Iland after him Gathelia, and after his wife Scotia.*"[519] Much the same statement about Ireland being named after Gathelus appears in the, 1571, work by Edmund Campion entitled *Historie of Ireland*.[520]

That Ireland was known as Gathelia is also corroborated by a statement in William Cambden's, seventeenth century, book *Britannia*. He states, with reference to the historic Scottish county of Argyll, that "*Beyond the Lake Lomund and the West part of Lennox, there spreadeth itself nere unto Dunbriton Forth the large country called Argathelia and Argadia in Latin, but commonly Argile, more truely Argathel and Ar-Gwithil, that is, Neere unto the Irish, or, as old writings have it, The Edge or border of Ireland, for it lieth toward Ireland.*"[521] (The old writings to which he refers probably includes the early 13th century *De Situ Albanie*.[522]) Similarly, in the Dean of Lismore's Book[523] (a collection of poetry from the 16th century) there is a reference to the lands on the west coast of Scotland as "Argathelia". The author wrote "*Somarled, the founder of the race, first appears in history as regulus or petty king of Oirirgaidheal, or, as it is given in the Irish form, Airergaidheal, and, in the Latin, Argathelia, a name which had sprung up subsequently to the ninth century, signifying the coast-lands of the Gael, and embracing the entire west coast from Cowall to Loch Broom.*"[524]

Returning to the matter of etymology, there are two important issues in these quotations to which attention should be drawn.

> ➤ In Cambden's book he states that the area is "*called Argathelia and Argadia*", thereby indicating that the "th" and "d" forms of the name are interchangeable and also (and possibly more importantly) that Ireland may (it being accepted that the 'Ar'

derives from a word indicating proximity or edge) have been known as "Gadia" or "Gathelia".

> Secondly, in the Dean of Lismore's book the Scottish and Irish Gaelic names for the area both use the 'd' (rather than 'th') form - *Oirirgaidheal* and *Airergaidheal*.

Finally, Cambden also refers to certain inhabitants of Scotland (who, it must be remembered, are said to have moved across from Ireland) as Gadeni. He states *"Upon the Ottadini or Northumberland[525], bordered as next neighbours, the* Γάδενοι, *that is, Gadeni, ..."* (Google Translate translates the Greek 'Γάδενοι' as either Gadenes or Gadenoi.)

It is therefore reasonable to accept, from the statement that *"they agreed to name* [Ireland] *after him Gathelia"* and the explanations given for the name of Argyle, that Ireland was probably, in times past, known as Gathelia or even (taking the *"Gaidel"* translation of Gathelus' name) as Gadia, Gadelia or Gadeira.

It therefore appears indisputable that Ireland previously had a name with a very close and distinct similarity to Plato's name Gadeira. (And that some the Scots, who were an offshoot of the Irish, were known as Gadeni/Gadenes.)

9.2.2.3 Is it chronologically possible that Plato could have known of Ireland as Gadeira?

If it is therefore accepted that it is reasonably possible that a reference to Gadeira could have been a reference to what is now known as Ireland then the question which arises is 'Is it possible, in terms of the chronology of events, that Plato could have known Ireland as Gadeira?'

In order to answer this question one must first know when the Milesians are said to have arrived in Ireland.

There are many different dates, with a wide span, proposed for the arrival of the Milesians in Ireland.

According to Keating's *General History of Ireland,*[526] *"the Milesians landed in the island about 1300 years before Christ was born ..."*. Martin Haverty, in his work *The History of Ireland: ancient and modern,*[527] states that *"It was in the year of the world 3500, and 1700 years before Christ, according to the Four Masters, or A.M.*[528] *2394, and B.C. 1015 according O'Flaherty's chronology,*[529] *that the Milesian colony arrived in Ireland."*[530] In *The history of Ireland* by Thomas Moore[531] it is stated that *".... the two sons of the great leader, Milesius, at length fitted out a great martial expedition, and set sail, in thirty ships, from the coastline of Gallicia to Ireland. According to Bardic chronology, 1300 years before the birth of Christ, but according to Nennius Aengus, and others, near five centuries later ... "*[532]

Taking these parameters into account it appears that the Milesians arrived in Ireland sometime between 1 700 BCE and 800 BCE.

Plato was born sometime between 428 and 423 BCE and died in 348/7 BCE[533] and, as such, it is possible that Ireland was known by the name Gadeira (or something similar) at the time that Plato was alive.

9.2.2.4   Gadeira and Erytheia

Gadeira was not, as was seen above, the original name of the island in question. It was previously known as Erytheia – or some similar spelling.[534] The current name in the Irish language is Eire, a name not entirely dissimilar to Erytheia – particularly in pronunciation in Irish and Greek respectively. This leads to the question as to how long the name Eire has been extant. Eire, as the name of the current Irish Republic, has only been so since 1937[535] however the word and name possibly dates back to the Proto-Goidelic language and even possibly to a Proto-Indo-European. The Proto-Indo-European word underlying the name Eire has a meaning suggestive of "abundant land".[536]

It should also be borne in mind that there was a non-Indo-European language spoken in Ireland before the arrival of the Indo-European / Goidelic peoples. It would appear that, after the arrival of the Goidelic peoples, these original Irish people were relegated to vassal status. Vestiges of their language might have lived on until the first millennium CE and their names might live on in certain place names.[537,538,539]

Geoffrey Keating in his book *The History of Ireland*[540] lists fourteen names which had been given to Ireland and, of the fourteen, Eire is the fourth.[541] If there is a link between the word Erytheia and Eire this may explain why is was known as Erytheia before it was known as Gadeira.

9.3   Conclusion

Given the above it appears to be probable that
> Gadeira (or Gades) and "Erytheia" are one and the same place and

> are, what is now known as, Ireland/Eire.

That being so, it is also probable that when Plato referred to the "*country which is now called the region of Gades*" he was referring to what is now known as Ireland - not the modern-day Cadiz as has generally been accepted.

9.4   Afterthought

As an aside it is also interesting to note, considering that
> according to the mythological story of the labours of Heracles, Heracles was required to bring back the red cattle (or kine) of Geryon who lived in, or near, Erytheia, and

> according to McGee's *A Popular History of Ireland,* the second invasion of Ireland was by people whose route to Ireland is described as being via "… *the Baltic Sea, and from the Baltic to Ireland*",

C.J. Edwards *et al.* in a study[542] of the origins of European dairy cattle found that the genetics of European dairy cattle reflect that the founders of the pied or red breeds originate from the North Sea and Baltic coastal areas (And the spotted, yellow or brown breeds from Switzerland).

I f there is a factual basis to the story of Heracles and to the invasions of Ireland, then this also, possibly, provides tenuous circumstantial corroboration for the proposition that Erytheia was Ireland.

I shall now move on to consider, in the next section, certain further circumstantial evidence which may corroborate the many links and congruencies between Plato's description of Atlantis and the known facts about Hatton Rockall.

SECTION TEN:
CIRCUMSTANTIAL EVIDENCE

10.1   Introduction

In this section I shall consider certain circumstantial evidence[543] which possibly substantiates certain aspects of Plato's description, namely:

(i)	the issue of elephants being on Atlantis,
(ii)	the minerals which occurred on, and near, Atlantis,
(iii)	certain genetic 'coincidences' and
(iv)	other 'coincidences' and anomalies.

10.2   Elephants

10.2.1   Reiteration

Plato states, in the *Critias*, that "*Moreover, there were a great number of elephants in the island; for as there was provision for all other sorts of animals, both for those which live in lakes and marshes and rivers, and also for those which live in mountains and on plains, so there was for the animal which is the largest and most voracious of all.*"

10.2.2   Elephants and Mammoths

Although there is a difference between elephants and mammoths the subtle distinction between elephants and mammoths (if it ever existed in ancient Greek) could well have been lost between the time of the actual events and their narration by Plato "9 000" years later. That said there were two proboscidean (elephant like) species which lived in Europe during the Pleistocene. These were the woolly mammoths (*Mammuthus primigenius*) and the straight-tusked elephants (*Paleoloxodon antiquus*). They occupied different environmental niches. The mammoths were more cold adapted, ate mainly grass and sedges and were primarily grazers. Straight tusked elephants, other hand, preferred warmer, temperate, conditions and flourished in the interglacial periods during the Pleistocene. They, it appears, preferred wooded environments and were predominantly browsers.[544,545] It is probable, given that straight-tusked elephant were found in temperate climates, that it is they who were found on a temperate Atlantis/Hatton Rockall. Apart from this it is apparent from the statement (in the Jowett translation) that *"There was an abundance of wood for carpenter's work, and sufficient maintenance for tame and wild animals"*[546] that Atlantis was well wooded. This would, then, also be consistent with the straight-tusked elephant having been found on Atlantis.

The issue then is, is it possible that straight-tusked elephants still existed at the time of the destruction of Atlantis?

The straight-tusked elephant is believed to have become extinct in the British Isles as early as 115 000 years ago however it did remain in existence in the warmer Iberian peninsula refuge until about 30 000 years ago.[547] There have also been suggestions that it did not die out until about 3 000 years ago in parts of China.[548]

Putting aside the, possibly questionable, survival of this elephant in China, is it is possible that it could have survived and remained in existence (beyond the time of the last known survivors in the Iberian Peninsula) in a refuge on a subaerial Hatton Rockall? [There is no doubt, as I set out below, that mammoths could, and did, survive well beyond Plato's date for the sinking of Atlantis.]

There is no direct evidence that straight-tusked elephants did survive on any other island refuge but, on the basis that 'the absence of evidence does not constitute proof of absence', one may consider the examples of other animals which are known to have survived in refuges beyond the commonly accepted extinction dates.

Before going further it is also worth noting that, according to Masseti[549] the straight-tusked elephant, *Palaeoloxodon* (or *Elephas*) *antiquus,* was found on the islands of the Tuscan Archipelago in the northern Tyrrhenian Sea. They probably arrived there during the Würm Glaciation which ended about 11 700 years ago[550] – at much the same time as Atlantis is said to have sunk.

That noted, we can now consider how certain other animals survived for extended time periods in refuges.

10.2.2.1   Animals surviving in refuges

The first animal to be considered is the mammoth.

Although it is generally considered that, in the United Kingdom and Ireland, the most recent of mammoths' remains are about 14 000[551] years old there are other more recent remains which have been found elsewhere. Remains found in Estonia date to about 12 000 years ago,[552] and the most recent Holocene mammoth remains in Europe are dated to ~9 760 years ago (Russia) or ~9 030 year ago (Finland).[553] Other, more recent, remains dating to approximately 5 900 BCE have been found in a refuge on St Paul Island off Alaska[554] and the most recent mammoth remains have been dated to 2 000 BCE,[555,556] - almost in historical times. The latter remains were found in a refuge on Wrangel Island in the Arctic Ocean. It is therefore indisputable that mammoths were able to, and did, survive, in certain island refuges, until well into the Holocene – and well after the date given for the flooding of Atlantis.

The second animal to be considered is the Irish Elk, or giant deer (*Megaloceros giganteus* Blum.).

The Irish Elk was generally believed to have gone extinct in Western Europe between approximately 12 600 to 12 700 years ago and in the British Isles and southern Scandinavia at about 11 500 years ago. In a 2015 article by Van der Plicht, J., *et al.*[557] it was concluded that these giant deer had, in fact, survived for a further, nearly, 4 000 to 5 000 years in a refuge in the Trans-Urals area – until approximately 7 900 to 7 600 years ago.

Given these examples (others may also be found) it is apparent, and uncontentious, that 'pockets' of animals can, given the right climatic and environmental conditions, survive for thousands of years beyond when most of their species have gone extinct.

10.2.3   Conclusion

It is, therefore, apparent that 'elephants' (whether mammoths or straight-tusked elephants) could, consistent with Plato's description, have roamed a sub-aerial Hatton Rockall island.

10.3 Minerals on Atlantis

10.3.1   Introduction and reiteration

This aspect is, I realise, particularly speculative and tenuous as the mere fact that a mineral exists in one area does not mean that it will occur in another many kilometres away. I have, nevertheless, chosen to include it because Plato did mention the mining of minerals and therefore (accepting, for the sake of this, some truth to what he describes) the minerals and metals must have been obtained from, at least, somewhere relatively close.

Plato states, in the *Critias*, that "*they dug out of the earth whatever was to be found there, solid as well as fusile, and that which is now only a name and was then something more than a name, orichalcum, was dug out of the earth in many parts of the island, being more precious in those days than anything except gold.*" He also states that "*The outermost of the walls was coated with brass, the second with tin, and the third, which was the wall of the citadel, flashed with the red light of orichalcum.*"

It does not, of course, follow that the mere fact that the walls were covered with these metals, that they all came from the island. For this reason I shall now consider whether it is possible, or likely, that tin and copper (being the metal of which brass is formed) could be found within the Hatton Rockall or surrounding areas. In order to do this (there being no information, of which I am aware, relating to whether or not there is, as a matter of fact, copper and tin to be found in Hatton Rockall) I shall, again, look at Ireland and England.

10.3.2   Ireland

Ireland, according to the website of the Eire Department of Communications, Energy and Natural Resources,[558] has a long association with mining – with records of mining dating back to at least 2 000 BCE. The south-west of Ireland was then an important copper producer. Alluvial gold was also found and used to produce gold artefacts. Today Ireland is still the largest producer of zinc in Europe and the second largest producer of lead.[559] Zinc is a common ingredient of brass.[560] Bronze, on the other hand, is another copper alloy which usually contains, tin[561]. Tin is not common in Ireland but its ore, cassiterite, (or tinstone) has been found in, or near, Wicklow, Dublin, and Killarney.[562]

10.3.3　England

"*Historically extensive tin and copper mining has occurred in Cornwall and Devon,* [England] *as well as arsenic, silver, zinc and a few other metals.*"[563] The mining for copper and tin started at least as early as 2 150 BCE. Because of the richness of the resources in Cornwall it remained one of the most important mining areas in Europe until the early 20th century.[564]

Herodotus, who I have mentioned before, wrote of the *Cassiterides*, or Tin Islands[565] which were regarded as being somewhere off the west coast of Europe. It has been proposed, because of the richness of the tin resources, that the British Isles are the Cassiterides.

The mining of tin in southern England was also described by Diodorus Siculus[566] in Book V of his *Library of History*.[567] He states "*The inhabitants of Britain who dwell about the promontory known as Belerium are especially hospitable to strangers and have adopted a civilized manner of life because of their intercourse with merchants of other peoples. They it is who work the tin, treating the bed which bears it in an ingenious manner. This bed, being like rock, contains earthy seams and in them the workers quarry the ore, which they then melt down and cleanse of its impurities. Then they work the tin into pieces the size of knuckle-bones.*"

10.3.4　Conclusion

It is apparent from this that, even if these minerals are not available on Hatton Rockall itself, they are to be found in relatively close proximity.

10.4　Genetic 'coincidences'

10.4.1　Introduction

It is stated in the *Timaeus*[568] that "*the empire of Atlantis reached in Europe to Tyrrhenia*[569] *and in Libya to Egypt*" and also "*Now in this island of Atlantis there was a great and wonderful empire which had rule over the whole island and several others, and … the men of Atlantis had subjected the parts of*

Libya within the columns of Heracles as far as Egypt, and of Europe as far as Tyrrhenia."

Accepting, for the moment,

➤ that there might be a factual basis to Plato's statement that the Atlanteans had an empire within Europe and had invaded, and conquered, large areas within the Mediterranean and

➤ human nature and mankind's tendency to leave offspring wherever they go,

one may consider whether it is possible that there might be some residual genetic traces?

Before dealing with that question and as an aside, it is interesting to note that there are a number of species, including the strawberry tree, the Kerry slug, and the Pyrenean glass snail, which are exclusive to Ireland and Iberia. The reason for this may, it has been suggested, be most easily explained by the movements of ancient people.

10.4.2   *European genetic outliers*

Now to consider the possible genetic traces.

The majority of the present western European population does not have roots extending back to the upper Palaeolithic[570] / early Holocene.[571] However, the Basques, whose historical territory lies on the border between France and Spain, are direct descendants of the Palaeolithic Europeans. This has been generally accepted and has also been established by genetics.[572,573]

The Basques, although they speak a non-Indo-European language, are also, in a 'principal components analysis' of gene-frequency data, most closely related to the peoples of the Atlantic fringe.[574] The Irish of western Ireland, and in particular those of Connaught, share a very high prevalence of the hg1 Y chromosome with the Basques. Frequencies of this gene show a cline across Europe with the lowest prevalence (1.8%) being found in Turkey and the highest peak, in mainland Europe, being amongst the Spanish Basque (89%). The cline continues into Ireland and reaches its highest prevalence in Connaught, western Ireland, where it reaches 98.3%.[575,576]

It is, therefore, apparent that (notwithstanding the fact that these Atlantic fringe peoples generally speak Celtic languages,[577]) there is a clear genetic link between the inhabitants of western Ireland and the Basques who are descendants of the original Palaeolithic inhabitants of Europe. This may also provide a degree of substantiation to Irish mythology which that states that the ancient Irish were descended from peoples who came from, or had settled in, Spain.[578]

Also, possibly, sharing a distant Palaeolithic common root with the Basques are the Saami of Finland. (The Saami, although speaking a Finno-Ugric language, are not genetically closely related to the surrounding speakers of related Uralic languages.) The mtDNA and Y chromosomal heritage of the Saami arise from European roots deriving from the Franco-Cantabrian region (the present Basque heartland) and can probably be traced back to a recolonisation of northern Europe after the last glacial maximum.[579,580] Also of interest is the fact that it has also been found that the Saami, of far northern Europe, and the Berbers of North Africa are genetically related, at least on the maternal line.[581] It is postulated that this link is derived from a shared link to the Franco-Cantabrian region.

The Sardinians (Sardinia lies on the western edge of the Tyrrhenian Sea.) are also, like the Basques, genetic outliers in Europe and also have roots going back to the Palaeolithic era.

10.4.3   Conclusion

The conventional, accepted, explanation for these links is, as set out above, that these people all share a common root in, or link to, a Franco-Cantabrian refuge. An alternate possibility (and I put it no higher than that) might be that they all share a link to a 'refuge' on an island Hatton Rockall. That a subaerial island Hatton Rockall could have been a refuge during the Pleistocene ice age might be corroborated by the fact that a portion of southern Ireland itself was a refuge during the Pleistocene.[582,583]

10.5 Other 'coincidences' and anomalies

10.5.1 Cultural links

It is not possible here to investigate all possible cultural links between various European peoples. That said (and given their possible genetic links as mentioned above), we may consider that, according to the author R Frank,[584] there are also deep cultural links between the Basques and Sardinians. These links relate to the position of bears in their cultures (past and present) and the possible previous religious reverence for them.

10.5.2 Linguistic links

There is still debate as to the area from which the Indo-Europeans originated but it is generally accepted that they moved from the east into Western Europe during the Neolithic and that, over a period of time, their language/s came to dominate in Western Europe. What is often overlooked is the fact that Europe was almost completely populated before the arrival of the Indo-European speaking peoples.[585,586] That being so one might also expect some cultural links between peoples.

Venneman[587] proposes that there was an original, substratum, European language which he designates 'Vasconic'. This Vasconic, or other substratum language, has been found to influence words in Indo-European languages, particularly in words describing places, rivers, plants and similar items which the incoming Indo-Europeans might have found when they arrived.[588589]

It has also been found that there are parallels between early Sardinian words and words in the Berber language.[590] If there are genetic links between the Berber and the descendants of Palaeolithic, or Mesolithic, Europeans, as seems to be suggested by the above, this would be a further, possible, indicator of such a link.

10.5.3 An anomaly – Genetic diversity in the Canary Islands

As a general proposition, the further away a given genetic population moves from its founder population the less genetic diversity is found in the population. This is seen, for example, in the general human population with there being greater genetic diversity amongst African populations than amongst the "outside of Africa" populations.[591]

It is against this background that one needs to consider the, ostensibly, anomalous situation that is found to have prevailed amongst the Guanches, the aboriginal inhabitants of the Canary Islands. The Canary Islands are islands which now belong to Spain and which are in the Atlantic off the West coast of Morocco, Africa.

It is generally thought that the original settlement of the Canary Islands was by people from Africa, probably of Berber origins. Where the anomaly arises is that the greatest genetic diversity was found to exist (when studying ancient Guanche genetic samples) in the outermost island of La Palma.[592] This, on the face of things, is not consistent with settlement from Africa but rather from some point further away in the Atlantic.

It has also been found that the Guanches possessed the U6b1 lineages, which are in the present day Canarian population, but is not found in African populations. Similarly, the phylogenetically closest ancestor found in Africa, the U6b, is not present in the Canary Islands.[593]

10.5.4   *Palaeo-European genetics*

Finally, as an aside and still breaking from my stated intention to not become involved in issues relating to the human aspect of the Atlantean story, the following might be interesting.

In the closing pages of the *Critias* it is stated, with regard to the original inhabitants (the descendants of the 'gods'), that "... *but when the divine portion began to fade away, and became diluted too often and too much with the mortal admixture, and the human nature got the upper hand, then they, being unable to bear their fortune, behaved unseemly, and to him who had an eye to see- grew visibly debased ...*".

From this it appears that (according to the story) (i) the original inhabitants started inter-marrying or breeding with, presumably, new incoming people and (ii) the culture, and behaviour, of the original inhabitants started to change and (iii) this occurred, or reached a peak, just before the destruction of Atlantis ~ 11 500 years ago.

A recent study[594] has found that

➢ between ~37,000 and ~14,000 years ago all the ancient European population "… *descended from a single founder population …*" and

➢ it was only during "*the major warming period after ~14,000 years ago, [that] a genetic component related to present day Near Easterners became widespread in Europe.*" and

➢ this time period[595] correlates archaeologically with cultural transitions in Western and Southern Europe.

The study might therefore corroborate Plato's story of the interbreeding of the original inhabitants with later arrivals.

10.6 Conclusion

In this section, Section 10, we have considered a number of issues which, of themselves, do not prove that Hatton Rockall was Atlantis. From them though, we can see that it is possible that there were elephants on a sub-aerial Hatton Rockall (as Plato stated with regard to Atlantis) and that the metals allegedly used in building on Atlantis were available (if not on Hatton Rockall) in close proximity. Also, and whilst I have generally attempted to steer away from the more human aspect of the story, it appears that there might be genetic, linguistic and cultural links between peoples who could have fallen within the area of influence of an imperial Atlantis.

SECTION ELEVEN: FLOODING AND EARTHQUAKES

11.1   Introduction

I have, I hope, put up persuasive evidence that Hatton Rockall has a very close resemblance to the description given by Plato of Atlantis. That being so, the mechanisms by which Atlantis is said to have disappeared, namely flood and earthquake, must be considered.

It appears to me to be improbable that mountains which *"were celebrated for their number and size and beauty"* would have flooded and disappeared within a single day. Accordingly, and in order for the story and debate to continue, I must supply a hypothesis as to why it is plausible that Hatton Rockall has sunken to a thousand metres or more below sea level whilst still remaining true to the story.

The precise mechanism of this submersion is beyond my expertise however I shall, in the next section, refer to various scientific works dealing with the possible mechanism/s for the submersion of land in the area. I concede that, to the best of my knowledge, there are no accepted geological theories supporting the proposition that Hatton Rockall subsided within the last 12 millennia but (if it is accepted that the descriptions of Atlantis accord with the known facts about Hatton Rockall) we then have either an incredible coincidence or the accepted facts and theories must be reconsidered and possibly reinterpreted. I propose that the latter be considered.

11.2   Reiteration, extraction of information and discussion

I reiterate the relevant portions of the story:

➤ *"Let me begin by observing first of all, that nine thousand was the sum of years which had elapsed since the war which was said to have taken place ... "*

➤ *"Many great deluges have taken place during the nine thousand years, for that is the number of years which have elapsed since the time of which I am speaking......";*

➤ Atlantis *"sunk by an earthquake, [and] became an impassable barrier of mud to voyagers sailing from hence to any part of the ocean. "*[596]

➤ *"But afterwards there occurred violent earthquakes and floods, and in a single day and night of misfortune all your warlike men in a body sank into the earth, and the island of Atlantis in like manner disappeared in the depths of the sea. For which reason the sea in those parts is impassable and impenetrable, because there is a shoal of mud in the way; and this was caused by the subsidence of the island. "*[597]

11.2.1   The 9 000 year period

There are writers who have suggested that this nine thousand year period should, in fact, be regarded as a reference to nine hundred years and also that Plato inflated many of his numbers by a factor of ten. The assumption that the events took place about 900 years prior to Plato is probably as a result of the fact that most Western people believe that it is impossible for unwritten information to have been passed down accurately over 9 000 years and that, consequently, this period of 9 000 years must be incorrect. I have shown, in section 2, that it was, in fact, possible, in pre-literate times, for such information to be accurately conveyed.

As I have set out in section 4.3 , had Plato simply wanted to convey that the event happened untold thousands of year ago he could have used the term 'myriad' which was used to convey, depending on the context in which it was used, that something was uncountable or infinite. Plato, when referring to the period of nine thousand years, was therefore not simply referring to a time untold ages ago but specifically chose to refer to that time period.

As such it is to that time period that we must look – although it should not, in my opinion, be taken as an exact period. There are a few reasons why I say the nine thousand years should not be taken as an exact period. Firstly, he chose to use a 'round' figure of nine thousand years. It is improbable that the telling of the story would have been told on the exact anniversary of such an event. Secondly, it is also improbable that it was intended to convey a precise period down to the year because of the inconsistencies in his writings. By way of example, in the *Timaeus* he refers to Athens having been founded nine thousand years ago but in the *Critias* he refers to the Athenians as having, nine thousand years prior to the recitation of the story, fought a war with "*those who dwelt outside the Pillars of Heracles*". Thirdly, nine thousand years is, in relation to current knowledge of human cultural history, an enormous period and is bound to be subject to some degree of inaccuracy.

I shall therefore focus (although not being bound to an exact figure of nine thousand years) on events which, modern science accepts, occurred at or near that time. It must also be borne in mind that even the dates supplied in scientific studies and articles can be subject to a degree of uncertainty and inaccuracy.

Having chosen (in order to be consistent with Plato's story) to focus on the time period, and events, around the date specified by Plato, I recognise that, over such immense periods of time, one's perception of events is foreshortened – it all happened 'back then'. It becomes subject to what Barber *et al.*,[598] refer to as the "perspective principle". That being so, I cannot exclude the possibility (albeit that I think that it is unlikely in the light of the following evidence of events around the 9 000 year time period) that the submersion of Atlantis occurred at the time of the so-called "8.2 kiloyear event"[599] when there was also massive flooding, volcanism and earthquakes.

11.2.1.1　In "a single day and night of misfortune"

One other aspect which bears focussing on at this stage is the issue as to how long the destruction of Atlantis took.

It is commonly accepted that the destruction took place "*in a single day and night of misfortune*". If, however, one re-reads the relevant section it states that the "*violent earthquakes and floods*" which resulted in the men of Athens sinking into the earth took place in a single day and night. It then goes on to state that Atlantis "*in* [a] *like manner* [sank and] *disappeared in the depths of the sea.*"

(The *Perseus Timaeus* translation of this section reads: "*But at a later time there occurred portentous earthquakes and floods, and one grievous day and night befell them, when the whole body of your warriors* [Athens' warriors] *was swallowed up by the earth, and the island of Atlantis in like manner was swallowed up by the sea and vanished;*" [Sections 25c and 25d.])

It is unclear from this statement whether the statement "*in like manner*" refers simply to the earthquakes and floods or to the earthquakes and floods having occurred within "*a single day and night*". Whilst I am willing to accept that it is probable that a flood, of the proportions which shall be considered below, could have destroyed large parts of Atlantis within a day and night it is, I think, improbable that the whole of an island of the size of Atlantis, mountains and all, would have subsided within a day and night.

I therefore propose that it would be reasonable to consider the possibility that the complete submersion of Atlantis / Hatton Rockall took an extended period, probably millennia (at least to reach the depths at which it now finds itself). This might find accord in Plato's statement, in the Jowett translation, that "*Many great deluges have taken place during the nine thousand years ...*"

I understand that even the sinking over millennia is extremely fast in geological terms but it is, I hope, conceivable that a re-evaluation of existing, or new, facts might support this possibility.

11.3 Secondary proposition (regarding the submersion)

My secondary proposition is therefore that Atlantis, having been an island during the latter part of the Pleistocene and a very early portion of the Holocene,[600] was initially inundated (or submerged) by a flood of cataclysmic proportions and then sank over millennia, to its present depths. (The related issues of eustasy and isostatic adjustment will be dealt in Section 13.)

George Sarantitis, possibly corroborating the initial aspect of my proposition, is of the view that *Timaeus* 25d should rather be interpreted as that Atlantis was 'inundated' and that the choice of the word sank by most translators is erroneous.[601] The Lamb translation of *Timaeus* 25d,[602] leaves open the possibility of an inundation of Atlantis. It reads that *"the island of Atlantis in like manner was swallowed up by the sea ..."*. (I do acknowledge, in all fairness, that there are others who would nevertheless continue with the translation that Atlantis 'sank'.) In any event, Plato, repeating an ancient story as he was, would have been unlikely to have specified, or probably even have known, the depth to which Atlantis was covered – once the city and low-lying land was submerged it was, to all practical intents and purposes, destroyed, sunken and gone.

There is no academic support, as far as I am aware, for my secondary proposition but I propose that it should be considered a logical possibility. The trigger/s of geological events can be many fold and, consequently, the precise order of historical (and, particularly, prehistoric) events is not something which can be easily determined. Whatever the order of events, the sinking of Atlantis was probably the result of a concatenation of cataclysmic events.

I shall now consider the events which occurred at around nine thousand years prior to Plato.

11.3.1   *The era*

I shall, when considering Plato's description of the destruction of Atlantis and comparable, scientifically accepted, events, confine myself to the period from 10 500 BCE[603] to 8 500 BCE[604], i.e. the period one thousand years before and one thousand years after Plato's stated date of Atlantis' demise.[605] The reason for this is that I accept that both the scientific and Plato's dates contain an element of uncertainty or inaccuracy.

This period was one of great environmental and climatic change and covers what is regarded as the beginning of the Holocene epoch. It was an era of global warming, the retreat of the northern ice caps, the creation of melt-water proglacial lakes,[606] truly cataclysmic floods and very powerful earthquakes.

11.3.2   The Palaeofloods

Many palaeolakes, impounded by ice dams, formed at various stages in both Northern America and Northern Eurasia. Sometimes these ice dams broke abruptly (not only during the period in question) releasing tens of thousands of cubic kilometres of water over a limited period.[607,608,609] It is these truly massive floods which we shall consider.

Before dealing with the specific palaeofloods referred to below I wish to give some comparisons with which to compare the palaeofloods and to, so, give some idea of the immensity of those floods.

Firstly, during 2007/8 an Antarctic a sub-glacial lake flooded releasing between 4.9 km^3 and 6.4 km^3 of water.[610] This, comparatively speaking, miniscule amount of water would have been sufficient to meet all of Australia's water needs for about 6 months.[611]

Secondly, one may consider that the 2011 tsunami, which wrought such devastation in Japan, had an initial volume of only 250 km3[612] which is also, when compared to the volumes of the palaeofloods considered below, small.

These ostensibly large and, in the case of the tsunami, very destructive volumes of water were, in fact, very small compared with the volumes of the floods considered below.

Before going further, another factor which needs to be emphasised is that before approximately 10 000 BCE,[613] the Bering Strait, as we know it today, was not open and Alaska and eastern Siberia were linked by a land 'bridge' known as Beringia.[614] The result of this is that the only, or primary, route through which that water could leave the Arctic Ocean was through the Fram Straits[615] into the, now, Norwegian Sea (the passage between Norway and Greenland and the Denmark Strait) and then on into the Atlantic.[616] Hatton Rockall lies at the southern exit of the Norwegian Sea. See Figure 11.1.

Figure 11.1 View of the world from above the Arctic. [617]

<u>KEY (to Figure 11.1)</u>
1. Bering Strait   2.   Approx. North Pole
3. Fram Strait   4.   Norwegian Sea
5. Hatton Rockall   6.   Mackenzie River mouth

In addition, at that time, sea levels were lower because of the amount of water 'locked up' in the ice caps and the palaeolakes. Keigwin *et al* record there was enhanced sea-ice export through the Fram Strait around the beginning of the Younger Dryas.[618]

I shall now move on to consider various potential candidates for a flood that could have flooded and destroyed Atlantis. When considering the various specific palaeofloods I shall focus on those floods which could be geographically relevant to my proposition.[619]

I shall deal with the issue of earthquakes in section 11.3.3.

11.3.2.1   Lake Agassiz

The first, and possibly most well-known, palaeoflood which could be relevant is the one that resulted from a catastrophic draining of the palaeolake Agassiz.[620] Lake Agassiz was, after the glacial maximum, the largest palaeolake in North America. Over time, it covered large parts Canada and the north west of the United States of America, including Manitoba, Saskatchewan, Ontario, North Dakota, and Minnesota. Mann *et al* have calculated[621] that, at two different stages of its existence, the area and volume of Lake Agassiz were,

- at approximately, 9 500 – 9 000 BCE[622], respectively ~152 500 km², and ~13 100 km³ and

- at approximately 7 900 – 7 500 BCE[623], respectively ~350 400 km² and ~38 700 km³.

Lake Superior, which is third largest lake in the world, has, by way of comparison, a volume of approximately 11 600 km³.[624]

There were a number of breaches of Lake Agassiz over its, approximately, 4 000 year lifespan. Teller *et al.*[625] provide a table of the volumes, fluxes, proposed routeing and timing of 10 different outbursts from Lake Agassiz. It is apparent from the table that estimated volumes, varying between 2 100 km³ and 163 000 km³, were released from time to time over its lifespan and that there were 4 outbursts over the period from *circa* 10 900 BCE to 8 600 BCE. The volumes of these outbursts ranged[626] from 9 500km³ to 7 000km³

There is some debate over the duration of these flood outbursts with some authors saying that each could have occurred over the course of about only one year[627,628] but this is doubted by Keigwin *et al*[629] who propose that it was longer. Whether it happened within one year or over a longer time period it appears undisputed that, considering the volume released, it was rapid and involved the release of massive volumes of fresh water into the Arctic Ocean.

According to Murton *et al.*[630]there were two catastrophic outburst floods[631] which flowed down, what is now, the Mackenzie River[632] in northern Canada and into the Arctic Ocean. The first of these floods occurred sometime between approximately 11 000 and 9 700 BCE.[633] This has been proposed as the possible cause[634] of the cold 'snap' known as the Younger Dryas[635]. The second was, according to the authors, between approximately 9 700 and 7 300 BCE.[636] This is also possibly corroborated by Keigwin *et al.*

Fisher, Smith and Andrews[637] hypothesise that another flood, beginning at about approximately 9 300 BCE,[638, 639] and lasting for about 1.5 to 3 years was the cause of the Preboreal Oscillation, a 'cold snap' lasting approximately 150 to 250 years after the climate had again started warming after the end of the Younger Dryas. The timing of this second flood accords well with the stated date for the flooding of Atlantis. The volume of the water discharged by this flood would, according to their estimation, have been a maximum of 21 000 km^3 which would have been equivalent to an initial 6 metre sea level rise in the Arctic Ocean (and, eventually, a 0.062 m rise in global sea levels). (This does not appear to be consistent with the estimated volumes referred above.)

There can be no doubt that, whether there was a flood of even a third of the magnitude proposed by Fisher *et al* or of any of the lesser volumes proposed by the other authors, the flood would have been devastating in the north Atlantic region and to a sub-aerial Hatton Rockall.

11.3.2.2   Siberian lakes

There are records of various outburst floods in Siberia but Soviet geological surveys and academic research studies have not been easily, if at all, accessible to international readers. Furthermore, the chronology of the events described is (from the information which I have had at my disposal) often uncertain or obscure.

That said, according to the authors[640] of the article *Glacial Lake Vitim*, a 3 000 km^3 outburst flood from Siberia to the Arctic Ocean', occurred at about 11 000 BCE.[641] A canyon (some 15 km from the remaining lake) which is 100 metres above the present-day River Vitim and unrelated to currently existing drainage patterns is a factor that leads them to think that there was cataclysmic outburst flood. The canyon is 6 km long, 2 km wide and about 300 metres deep.

According to Lioubimtseva *et al* another massive Siberian palaeolake might have existed, covering most of the West Siberian Plain at around the time of the Last Glacial Maximum. It stretched about 1 500km from north to south, and a similar distance east to west at its widest. It could also have drained into the Arctic Ocean.[642] This lake would, had it existed today, probably have been classified as a sea. At its maximum extent it had had a surface area of at least twice that of the Caspian Sea.

Palaeolakes held back by ice dams repeatedly formed, according to Reuther, Anne U., *et al.*,[643] during the Pleistocene in the Russian Altai Mountains. When these ice dams failed, the resulting cataclysmic floods caused resulted in some of the world's largest flood waves ever documented.

11.3.2.3 Baltic Ice Lake

There was, at least, one ice lake in the area of what is now the Baltic Sea[644]. This Baltic Ice Lake catastrophically drained approximately 18% of its surface area at the end of the Younger Dryas.[645] Jakobsson *et al.*[646] concluded that a volume of approximately 7 800km^3 of water drained during this event. This would have drained into what is now the North Sea. This drainage would, according the authors of the respective articles,[647] have taken place at a similar time to the drainage of Lake Agassiz; as studied by Fisher *et al.*.[648]

(Whilst it is beyond my knowledge and expertise, it does appear to me that this, possible, double drainage could explain why Fisher *et al.* calculate a significantly larger drainage from Lake Agassiz, at that time, than that calculated by Murton *et al.*[649] in their article '*Identification of Younger Dryas outburst flood path from Lake Agassiz to the Arctic Ocean.*')

During the Younger Dryas, parts of western Norway, unlike other areas in Scandinavia, experienced a sea level rise of about 10 metres.[650]

11.3.2.4 Conclusion

Considering that all, or at least most, of the flood waters considered above would have had to pass the Hatton Rockall area, the volume and probable height of the flood waters and the fact that the 'city' of Atlantis and surrounding plain must have been low lying, the Atlantean settlements on the plain would have been flooded dramatically and very quickly. Floods such as those considered above would also have carried an immense amount of silt and other debris[651] – including trees and other floating debris. This floating and suspended debris and silt would cause the sea in the area to be impassable to ships and boats - particularly primitive craft such as may have existed at the time.

This rapid flooding and rendering of the sea impassable coincides with the descriptions given by Plato.

11.3.3 Earthquakes

Having considered the immense floods which modern science accepts took place around the time given by Plato for the destruction of Atlantis we now need to consider whether it is also accepted, or likely, that there were also earthquakes at that time.

11.3.3.1 Can there be links between floods and earthquakes?

First we can consider whether there may be any link between water movement and earthquakes.

'Reservoir-induced seismicity' is a recognised phenomenon[652] which has generally been observed, and studied, in relation to the filling of dams. (It has even been suggested that, in certain circumstance, exceptional rainfall can induce seismic activity.[653]) But it is not only the filling of a dam ("loading") but also its emptying ("unloading")[654] that may give rise to seismic activity.[655,656] Most recorded reservoir-induced seismicity has resulted in relatively mild earthquakes but there have been a number of reservoir induced earthquakes registering in excess of M5 on the Richter Scale.[657] It is worth considering, at this stage, that the largest volume dam in the world, Lake Kariba, has a nominal volume of 'only' 180 km^3.[658] According to Wikipedia it is believed that, when it was filling, it caused induced seismicity in the region, including over 20 earthquakes of greater than 5 magnitude on the Richter scale.[659] Likewise in 2016 (when its level was down to only 12% due to drought[660]) there was an earthquake of close to magnitude 5 on the Richter scale.[661]

Apart from the issue of earthquakes due to the emptying of the palaeolakes, according to Brothers *et al*[662] the rapid late Pleistocene / early Holocene sea-level rise could have been sufficient to cause stress failures across the fault systems of passive continental margins and have triggered fault reactivation and rupture.

It is, therefore, logical (particularly considering the massive volumes of water released by the flooding) to assume that there must have been earthquakes of a very significant magnitude when the palaeolakes breached and flooded. We shall investigate this further in the next section.

11.3.3.2   Is there any evidence for earthquakes of very significant magnitude when, and where, the palaeolakes flooded?

After the end of the last glacial maximum and as the glaciers melted and palaeolakes disappeared, many parts of the world which had been covered by glaciers underwent a notable increase in seismicity. Scandinavia, parts of North America and even areas of the European Alps experienced this surge of seismic activity.[663,664] This, according to Hampel *et al.*,[665] was primarily due to the glacial melting and unloading. In Fennoscandia there were particularly powerful and significant earthquakes.[666] The largest of these earthquakes had a magnitude of at least Mw 8.2.[667,668] and had the fault lengths of up to 155 kilometres long and vertical displacements of up to 15 meters. Prominent examples of these faults and scarps, of which there are a number in Scandinavia, include the Pärvie fault in northern Sweden and the Stuoragurra fault in Norway.[669] The Pärvie fault is the largest known postglacial fault in the world. It is 155 km long, trending north northeast. The land along the length of the fault was vertically displaced (creating linear scarps) which average between 10 to 15 metres high[670] - this is the height of an average 5 storey building. This was probably[671] created in one cataclysmic event[672] sometime between about 9 500 to 7 500 BCE.[673,674] The Stuoragurra fault (and scarps) trends similarly north easterly. It is approximately 80 km long and has a displacement of about 10 metres.[675] According to Dehls, *et al.*,[676] there appear to have been earthquakes of magnitude 6 or more, during two periods, namely one before about 9 000 BCE[677] and another between 8 000 and 7 500 BCE[678]. To give some sense of perspective of the power of such earthquakes – the single 2011 Tōhoku, Japan earthquake, which had a magnitude of about 9,[679] is said to have moved the main island of Japan (Honshu) 2.4 m to the east and shifted the Earth on its axis by estimates of between 10 cm and 25 cm.[680] According to Sutinen *et al.*,[681] Northern Fennoscandia also experienced high-magnitude postglacial fault events. After studying the geomorphology of the area (including various structures formed during the discharge of subglacial lake(s)), they conclude posit that the event occurred at approximately 9 900 BCE.[682]

Hampel *et al.* also comment that the post glacial increased seismicity would have affected areas covered by smaller ice caps (or resultant palaeolakes) and would not have been restricted to the regions of the large Laurentide and Fennoscandian ice sheets.

Given the above it is apparent that, around the 9000 year marked given for the sinking of Atlantis, Europe, and probably North America,[683] were experiencing massive earthquakes.

11.3.3.3   Subsidence of the earth's surface due to earthquakes.

Large earthquakes can cause various degrees of subsidence of the ground over large areas, even up to 200 km from the rupture area. This was evident in the 2011 *M*w 9.0 Tohoku, Japan and the 2010 *M*w 8.8 Maule, Chile earthquakes in which the areas of subsidence measured up to 15 to 20 km across.[684] The depressions are caused, it is hypothesised, by the deformation and subsidence of magmatic and hot plutonic bodies beneath the area.[685] This relatively small degree of subsidence is, obviously, insufficient to account for the sinking of Atlantis but it may have been one of many factors contributing to its demise.

11.3.3.4   Is there any recognised link between earthquakes and volcanism?

Whilst Plato does not specifically mention volcanoes it is worthwhile considering this as a further indicator of the times, environment and climate. Are there links between earthquakes and volcanism and was there any significant volcanism at the time?

Walter,[686] in his article '*How a tectonic earthquake may wake up volcanoes: Stress transfer during the 1996 earthquake–eruption sequence at the Karymsky Volcanic Group, Kamchatka*' cross references various other articles showing that many volcanic eruptions are preceded by earthquakes. According to the articles reviewed by him the Mount Pinatunbo eruption was preceded by a magnitude 7.8 earthquake in July 1990 it, an earthquake in the Andes in July 1995 preceded surface deformations at various volcanoes which had not been listed as active and Mount Vesuvius had its largest eruptions shortly after earthquakes in the Appenines. He concludes that the twin-eruption at the volcanoes Karymsky and Akademia Nauk were triggered by the release of certain fractures and consequent earthquakes. Similarly, Marzocchi,[687] in his article '*Remote seismic influence on large explosive eruptions*', found that there was a significant correlation between major volcanic explosive eruptions and earthquakes which had occurred (up to 1 000 km distant) prior to the eruption.[688] Also indicating a link between earthquakes and magmatic action was that fact that in 2002, after the M7.9 Denali fault (Alaska) earthquake, clear changes in geyser activity and a series of local earthquake swarms were noted in the Yellowstone National Park area. These resultant changes occurred despite the fact that the area is about 3 100 km from the epicentre of the earthquake. Some of the geysers experienced changed eruption frequency within hours after the arrival of the surface waves from the earthquake.[689]

In Chile, on the 22nd of May 1960, the most powerful earthquake ever recorded (with a magnitude of 9.5) struck. On the 24th of May 1960, 38 hours after the main shock of the earthquake, the volcano Cordón Caulle began a rhyodacitic fissure eruption which continued until the 22nd of July 1960.[690] The existence of a direct link between these was not specifically studied but it appears, in the light of more recent studies, that there was a link. Subsequent studies in Chile have found that there is a significant linking between volcanic eruption rates and earthquakes of $M_W > 8.$ [691,692] In 2006 Manga *et al.*,[693] in their article *Seismic triggering of eruptions in the far field: volcanoes and geysers* concluded that, as a general principle, earthquakes can trigger volcanic eruptions but that this is not invariable and they do not know how this occurs.[694]

Importantly, and approaching the issue from the opposite perspective, Wauthier *et al.*[695] concluded that a magmatic dike intrusion and eruption during January 2002 probably induced the largest magnitude earthquake recorded in the Lake Kivu, Congo area. This earthquake occurred approximately nine months after the eruption, on 24 October 2002.

The answer to the question as to whether or not there is any link between earthquakes and volcanism must, therefore, be 'Yes'.

This then leads us, knowing that there had been massive earthquakes at around the time that, to the next aspect.

11.3.3.5 Was there any significant volcanism around the time that Atlantis is said to have sunk?

Zielinski *et al.*[696] (1996) state that the longest, continuous, period of enhanced volcanism within the last 110 000 years occurred between 15 000 and 4 000 BCE[697] with the greatest concentration of large volcanic signals appearing between 11 000 and 5 000 BCE.[698] In an article[699] published in 2000 Zielinski reiterated that the greatest period of volcanic activity was between 5 000 and 11 000 BCE. The stated date for the sinking of Atlantis therefore falls solidly within the period of the greatest volcanism since 11 000 BCE (and, in fact, within the last 110 000 years). In 2016 another study[700] also concluded that there was a massive increase in volcanic activity at the end of the last ice age, caused by a combination of erosion and the melting of the ice caps. It is stated, in the abstract of the article, that *"continental lithospheric unloading due to ice melting during the transition to interglacials leads to increased continental magmatic, volcanic and degassing activity"*.[701]

In the 1996, Zielinski *et al.*,[702] article the authors commented that the magnitude of the signal of many of the eruptions over the relevant period (i.e. between 110 000 and 7 000 years ago) are significantly larger than that of the eruption of Tambora, in 1815 CE. The Tambora eruption was one of the most powerful in recorded history and, because of all the volcanic ash thrown up into the atmosphere, the year following it, 1816, became known as the 'Year without a Summer'.[703]

11.4 Conclusion

One may therefore conclude that, at the time that Plato said Atlantis was destroyed and consistent with his descriptions,

> ➤ there was extensive flooding, of magnitudes unknown in historical times, with concomitant clogging of the sea and

> ➤ this was accompanied by massive earthquakes - probably preceded, or followed, by powerful volcanic eruptions.

<h1 style="text-align:center">SECTION TWELVE:
CONGRUENCIES SUMMARISED</h1>

12.1 Introduction

Before moving on to speculate as to how Hatton Rockall sank to its present depth I shall summarise the congruencies, which we have considered, between Atlantis and Hatton Rockall. It is then for you, the reader, to decide whether the evidence that I have advanced persuades you that Plato's description of the island of Atlantis was probably true.

12.2 Table of congruencies

Item No.	Descriptions of Atlantis	Hatton Rockall
1	Atlantis was a large island	Hatton Rockall would have been a large island
2	Atlantis was outside the Pillars of Heracles and in the Atlantic ocean	Hatton Rockall is in the Atlantic Ocean
3	Atlantis was sufficiently far north to experience cold winds but was protected from them by the mountains	The north of Hatton Rockall is at about 59^{0}N and has a range of mountains at its north.
4	Atlantis had a central plain	Hatton Rockall has a central 'plain'.

Item No.	Descriptions of Atlantis	Hatton Rockall
5	Atlantis' plain "was smooth and even, and of an oblong shape … "	Hatton Rockall's plain is comparatively 'smooth' and has few natural gradients on it.
6	Atlantis' plain was large and largely rectangular	Hatton Rockall's plain is large and its length is longer than its breadth.
7	Atlantis' plain had "*a trench dug round about it*" the "*breadth and length* [of which] … *seems incredible…* ."	There is a natural trench around a substantial portion (if not all) of the edge of Hatton Rockall's plain which can be up to about 1 300 metres wide.
8	Across Atlantis' plain were cut "… *straight canals of a hundred feet in width* … [which] … *were at intervals of a hundred stadia.*"	Hatton Rockall's plain is criss-crossed by polygonal faults dividing the surface into sections and creating a honeycomb appearance.
9	Around Atlantis' plain the "*whole country was said … to be very lofty and precipitous on the side of the sea*"	Hatton Rockall's plain is bordered, on three sides, by tall mountains which are precipitous on the seaward side of the mountains.
10	In the middle of the island there was a small mountain.	Mammal Mount is a small mountain, which (when measured from east to west) is in the middle of the central plain.

Item No.	Descriptions of Atlantis	Hatton Rockall
11	This small mountain was surrounded by moats	Mammal Mount has, at least, one natural moat surrounding it.
12	Atlantis received most of its rain in winter.	Hatton Rockall receives most of its rain in winter.
13	Atlantis had an idyllic or temperate climate with luxuriant growth.	Hatton Rockall as an island most probably had a temperate climate with luxuriant growth similar to Ireland and Cornwall.
14	Atlantis faced a region called Gadeira (or Gades)	Hatton Rockall faces Ireland which has been known as Gathelia and the Irish were formerly known as Gadelians
15	Atlantis had "... fountains, one of cold and another of hot water ... "	Hatton Rockall had volcanic origins and would, most probably, have had hot springs.
16	Atlantis was destroyed by flood and earthquake around 9 500 BCE	There were massive floods in and around the Arctic Ocean and earthquakes in Scandinavia and surrounding areas at around 9 500 BCE
17	The flood and destruction of Atlantis rendered the Atlantic Ocean in that region impassable	The massive floods would, in all probability have carried a massive amount of debris and silt and would probably have rendered the sea impassable.

12.3   Conclusion

If, having considered all the information that I have provided, you are satisfied that

> ➢ there is a close correlation between the description given by Plato of the physical characteristics of Atlantis and the actual physical geography of Hatton Rockall and

> ➢ it is improbable that this degree of matching descriptions is purely by chance and, accordingly

> ➢ it is more probable than not that Plato, when describing the physical aspects of Atlantis, was describing what is now known as Hatton Rockall

then I have succeeded in my original stated intention.

12.4   The last issue – how Hatton Rockall sank

There is (accepting that Hatton Rockall was an island around the time of Atlantis' demise) one, last, remaining issue. This is 'How did it sink to its present depth in such a short geological time?'

This is a question which is well beyond my area of expertise but I shall nevertheless attempt, in the next section, to list factors and mechanisms which I propose could, as a matter of logic, have contributed to its subsidence. Whether or not these proposals are plausible will have to be answered by the appropriate experts.

SECTION THIRTEEN:
A POSSIBLE EXPLANATION FOR THE
CONTINUED SUBMERSION

13.1 Introduction

Once I had decided that, for me, the congruencies between Atlantis and Hatton Rockall went beyond mere random coincidence I wanted to find out if there was any explanation (which appeared plausible to me) for Hatton Rockall no longer being an island. The following are the results of my enquiries.

This section is not 'required reading' but, if you wish to join me in speculating about the causes and mechanisms of Hatton Rockall's subsidence, read on. I shall refer to a number of geological terms and concepts in the course of the discussions (primarily in section 13.6) however, in section 13.5, I provide explanations of the relevant geological terms and concepts.

Before considering these causes and mechanisms I shall, firstly, set out my proposition, then show that there are precedents for the subsidence and movement of large areas of land (in sections 13.3 and 13.4) and then, in section 13.6, set out, in some detail, the factors which I speculate could have led to the subsidence of Hatton Rockall. In support of my proposals I shall refer to various geological studies, theories and related facts.

13.2 Secondary proposition and introduction

To reiterate, the concept of Atlantis/Hatton Rockall sinking, to the depths at which it is now found, within one day appears totally implausible. (That is not to deny that it may have flooded within one day.) It is on the basis that the whole of Atlantis / Hatton Rockall did not sink in its entirety within one day, but rather took millennia to sink to its present level, that I advance the proposals in this section. I do not suggest that this subsidence was due to any one factor but rather that it was, in all probability, caused by the joint effect of various mechanisms which all occurred around the beginning of the Holocene. These likely mechanisms, discussed in section 13.6, include eustatic sea level rise, isostatic adjustment, possible reduction of magma bodies and also the compounding effect, on the subsidence, caused by the added weight of the flood waters.

13.3 Required rate of submersion and similar examples

For Hatton Rockall to take even 12 000 years to sink to its present position is extremely fast by general geological time frames. Rockall Islet, which would have formed the highest point on the island, stands approximately 1500 metres above the plain.[704] In order for it to have reached its present position (just above sea level) within 10 000 to 12 000 years it would have had to have submerged at an average rate of, approximately, 15 cm per year. This, being an average rate, is a very 'rough and ready' figure. It does not take into account many possible factors including, in particular, the possibility that the sea bed and island might have slumped initially at a much faster rate and then thereafter slowly subsided further - or vice versa.

There is still much uncertainty as to how the earth reacted to the ice masses and, particularly important for this discussion, the melting of the icecaps. This uncertainty is largely due to incomplete knowledge of, amongst other things, the earth's rheological properties over time.[705] We do know that the ocean bottom is still undergoing deformation due to increasing sea levels and melting glaciers. Frederikse *et al*[706] studied this present, ongoing, ocean bottom elastic deformation caused by present-day mass distribution. They found that ocean bottom deformation varies spatially and regionally.[707] Simply put, this means that the changes in the sea bottom vary from place to place – what might apply in one area may not be found in another.

Although this proposed rate of submersion is very fast in geological time frames, there are precedents for

> ➢ the subsidence of a large areas[708] of land at such a speed and

> the movement of a sub-continental (and continental) land mass at such a speed.

13.3.1   *Rapid subsidence of large areas*

There are two areas, in particular, which merit mentioning here. The subsidence of these two areas has been caused by humankind by the removal of an underlying, supporting, liquid.

The first area is that of the San Joaquin Valley, California (This is not a phenomenon which is restricted to the San Joaquin Valley. It has been observed worldwide wherever there has been substantial depletion of groundwater resources.[709]) The subsidence in the San Joaquin Valley started in about 1920 and by 1960 there had been a maximum subsidence of up to 20 feet (approx. 6 m). The area of subsidence covered 2 000 square miles (or approximately 5 180 km²) at that time[710] but in the next twenty years (by 1980) the area of subsidence more than doubled and affected nearly 5 200 square miles (or 13 470 km²). In addition to doubling the area of subsidence the depth of maximum subsidence had increased to up to nearly 30 feet (approx. 9 m).[711] The effect of this over 60 years is that the average rate of subsidence in the areas of maximum subsidence was about 15 cm per year – the required rate for Hatton Rockall to have subsided. In the last few years the rate at which the land has been subsiding has increased still further. In approximately nine months (from May 2014 to January 2015) a maximum subsidence of over 30 cm was found just south east of Corcoran in the Tulare Basin and near the California Aqueduct a maximum subsidence of, also, just over 30 cm occurred within a period of about four months (from June to October 2014).[712] This latter rate of subsidence (although it may not be sustained) amounts to a subsidence rate of in the order of 90 cm per year – larger by a factor of six than that required for Hatton Rockall to have sunk to its present depths over 10 000 years.

The second area of rapid subsidence is that of around the town of Wink in Texas, U.S.A. Kim *et al.*[713] studied this area (which covers approximately 4 000 km²[714]). The rate of subsidence has gone as high as 40 cm per year with 10 cm per year also being relatively common. The underlying causes of the subsidence are all, directly or indirectly, related to the removal of underlying supporting 'structures' - oil, gas and the like.

The examples of rapid subsidence in California and Texas have been caused by the withdrawal of the underlying and supporting groundwater, oil or other structures.

More recently, from about May 2018, the area around the island of Mayotte in the Indian Ocean has experienced a number of earthquake swarms, including one earthquake of magnitude 5.8. In addition there was a strange, unexplained, seismic event, on 11 November 2018, which was picked up by earthquake sensors stationed across the globe.[715] It has been hypothesised that these earthquakes and land movements are due to the emptying of a magmatic reservoir – which, it is estimated, is 37 km below the surface and 32 km to the east of Mayotte.[716] One of the results of these events is that the island of Mayotte has moved by about 9 cm to the east and has sunk by about 8 cm. What is particularly significant about this, for my proposal, is that the land surface has, in about 6 months, subsided by about 8 cm and that this subsidence was probably due to the movement away of magma. If this rate continues the rate of subsidence will exceed 15 cm per year and will be due, as far as is known, to a purely natural event.

In section 13.6 I shall extrapolate from these and consider the reduction of portions of the underlying, supporting asthenosphere of the mantle[717] and magma as a possible contributing factor to the subsidence of Hatton Rockall. In section 13.5 I shall deal, in general terms, with various geological terms and concepts.

Although not specifically dealing with subsiding land but rather with land movement the following two studies might also be of some relevance.

Barletta *et al.*,[718] in a 2018 study, showed that the Amundsen Sea Embayment area is presently rising at a rate of just over 4 cm per year. Whilst this deals with rising, rather than subsiding, land it is important because it also demonstrates possible rates of land movement. The authors predict that in a hundred years the uplift rates will be up to 3.5 times faster than at present. The uplift rate would then, in 100 years, be approximately 15 cm per year.

In 2016 *M.J. Hoggard et al.* found[719] that wave-like fluctuations, due to temperature-driven density anomalies or convection currents, occur in the mantle and, consequently, on the surface of the earth. These 'waves', the authors of the article found, have an amplitude of approximately 1 km, with wavelengths of approximately 1000 km's. This, according to the authors, is a much smaller areas than had previously been predicted. The action of these waves (i.e. the rising of the land surface) contributes to the building of mountain ranges and the subsidence of the sea bottom. They could also, I suggest, contribute to the submersion of islands. These 'waves' also occur at a rate that is much faster (an order of magnitude faster) than had been previously predicted. Hoggard, a co-author, commented[720] that "*Although we're talking about timescales that seem incredibly long to you or me, in geological terms, the Earth's surface bobs up and down like a yo-yo.*" and that "*These results will have wider reaching implications, such as how we map the circulation of the world's oceans in the past, which are affected by how quickly the sea floor is moving up and down and blocking the path of water currents.*"

Finally, and although it does not specifically relate to subsidence but rather to questions of climate, it is interesting to note Hoggard *et al.* also explained[721] that "*There are these narrow channels around Iceland that allow water* [the Gulf Stream] *to sink, ... If you elevate or depress them, you could really affect ocean circulation.*" Large-scale ocean circulation patterns move heat around the world and, if they are changed, they would also affect and change climates.

13.3.2   *Rapid movement of a sub-continental and continental land masses*

Regarding the second issue, India, before it joined with the Eurasian continent millions of years ago, moved north at a rate of between 15 cm[722] and 21 cm[723] per year.[724] This rate of movement, albeit lateral, is also within the required parameters and was a movement not induced, nor affected, by the actions of humankind. Similarly the entire Australian continent is currently moving northwards at about 7 cm per year.[725] The effect of this is that Australia has moved about 1.5 m (or nearly 5 foot) since 1994. Although this speed does not meet the requirements of my hypothesis it does show that such, relatively fast, movement is still occurring.

13.3.3   Conclusion

It is therefore apparent that large areas of land can move (both vertically and horizontally) at rates equal to, or greater than, that which would have been required for Hatton Rockall to reach its present depth within 12 000 years.

Before moving on to set out the factors, in section 13.6, which I propose could have led to Hatton Rockall's rapid subsidence I shall

> ➤ briefly discuss, in section 13.4, whether there is precedent for large areas of land subsiding beneath, or rising from, the sea to the extent contemplated for Hatton Rockall and, more particularly, whether there are any indications this may be happening in the Hatton Rockall area and

> ➤ give a short explanation, in section 13.5, of the geological terms and concepts which will be referred to in section 13.6.

13.4   Is there any precedent for land rising from and/or subsiding beneath the sea?

Charles Darwin, in his journal[726] entry dated the 29th of February 1832, wrote "*Along the whole coast of Brazil, for a length of at least 2000 miles, and certainly for a considerable space inland, wherever solid rock occurs, it belongs to a granitic formation. The circumstance of this enormous area being constituted of materials which most geologists believe to have been crystallised when heated under pressure, gives rise to many curious reflections. Was this effect produced beneath the depths of a profound ocean?* Darwin clearly suspected that this massive area, which is now dry land, had been formed beneath the ocean. Nearly 200 years later, Japsen *et al.*[727] studied the area and concluded that the region has, over millions of years, been subjected to multiple episodes of burial, uplift and erosion. The area is a classified as passive continental margin,[728] one of a number of sub-categories of continental margins. Passive continental margins may, in turn, be sub-divided into various types, one of which is the volcanic passive margin.[729]

In addition to simple passive continental margins there are also elevated passive continental margins ("EPCM"). These are, in essence passive continental margins which have been raised, or elevated, to form mountains. Examples of such elevated passive continental margins are the Scandinavian Mountains, Eastern Greenland, the Brazilian Highlands and Australia's Great Dividing Range. Japsen *et al.*, in an earlier work, [730] studied various other EPCMs and concluded that these areas had, similarly, been subject to episodes of burial and exhumation. They, referring to other published evidence (dealing with a number of margins), showed that periodic movements of the earth's surface, involving kilometre-scale burial and exhumation are common features of EPCMs.

Hatton Rockall, similarly, falls within a volcanic passive margin (See Figure 13.1)[731] but, unlike the Brazilian Highlands, it is not, currently, elevated. I emphasise, at this point, that the elevation of these areas has been on kilometre-scale.

This periodic subsidence and burial and uplift is also known to have occurred in other volcanic passive margins in the North Atlantic, including off Norway and Greenland[732,733] - these are both, geographically, close to Hatton Rockall. In Greenland the uplift is such that volcanic sequences, deposited during post-rift subsidence, are exposed in mountains reaching up to 2 km above sea level while certain marine deposits are found at elevations up to 1.2 km above sea level.[734,735]

It is, therefore, apparent that uplift can be large and significant and one would, concomitantly, expect that the subsidence could be similarly large. This is important given that Hatton Rockall appears to have sunk by about 1.5 km's.

Whilst there is there is no widely accepted geophysical model that explains the movement of the earth in these areas, it is probable that the underlying cause and mechanism of any uplift or subsidence is affected by the nature of the passive margin where it occurs. Hartley *et al.*,[736] concluded that the mechanism behind episodic uplift, and subsidence, in areas of the North Atlantic is changes in the underlying Icelandic mantle plume. The Icelandic plume extends, as I shall discuss later, under Hatton Rockall.

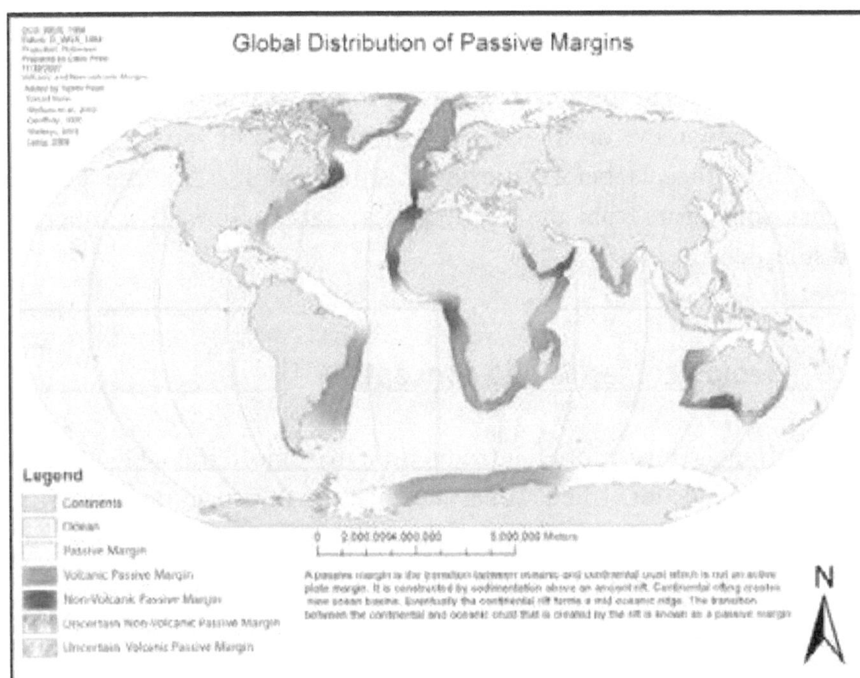

Figure 13.1 Map showing passive continental margins[737]

That the Rockall area (and, likewise, the nearby Faeroe–Shetland and Vøring areas) has undergone subsidence appears uncontentious. This was studied by Ceramicola, S. *et al.*[738] who concluded that there are kilometre-scale anomalies between the actual subsidence, based upon geological evidence, and the restored palaeo-depths[739] and that, notwithstanding the effects of uplift, there has been, overall, greater post-rift subsidence than rise along the North West European margin.[740] These anomalies are also present[741] in both the northern and southern Rockall Basin.[742,743] Zhirov[744] also proposes (at pages 298 to 303) that various parts of the North Atlantic were subaerial in the past and have now submerged. According to him the extent of their submersion was greater than the eustatic sea level rise and involved actual subsidence. This subsidence was, he says, a result of a tectonic process which *"ended geologically very late."*[745] Zhirov's proposal is based upon studies by various other Russian authors to whom I have not had access.

From this we may conclude, firstly, that it is possible for large areas of land to subside beneath the sea and, secondly, that there are subsidence related anomalies in the greater Rockall geographical area.

I shall now move on to give a brief explanation of various geological terms and concepts before considering, in section 13.6, the possible mechanisms (apart from the flooding) for Hatton Rockall's submersion and subsidence.

13.5 Geological terms and concepts

This section is a very brief introduction to some terms and concepts referred to in section 13.6. It does not take the investigation further but simply makes it easier for readers who, like me when I started on this journey, do not know these terms and concepts.

13.5.1   The earth's structure, generally

The earth may be divided, for geological purposes, into the crust, the mantle and the core. The crust is a thin shell on the outside of the Earth, accounting for less than 1% of Earth's volume, it is the top component of lithosphere which is the upper part of the mantle. The crust, mantle and core differ according to their chemical and physical (rheological) properties. The outer solid silicate crust is underlain by a highly viscous, almost solid, mantle. The mantle is divided into further sub-divisions – discussed in paragraphs 13.5.2 and 13.5.3. Beneath the mantle is the core which is comprised of an upper, extremely low viscosity, liquid outer core and an inner, solid, core. The core also has further subdivisions but these are irrelevant to this discussion.

13.5.2   The lithosphere

The crust and the relatively cold, rigid, top of the upper mantle are collectively known as the lithosphere (from the Greek word meaning 'stone'). The lithosphere is considered to extend to the depth of the isotherm associated with the transition between brittle and viscous behaviour, generally taken to be at about the 1 000°C isotherm.[746] The thickness of the crust varies from an average 6 km for ocean crust to between 30 and 50 km for continental crust.[747] This, I emphasise, is only an average and it varies from place to place depending upon circumstances prevailing in any one area. Beneath the lithosphere is the asthenosphere (to which I shall refer below), a relatively low-viscosity layer on which the lithosphere rides.[748]

The lithosphere generally remains rigid for very long periods of geologic time (during which it may deform elastically and through brittle failure).[749] It is a result of the lithosphere's inability to deform plastically that one finds faults forming and, after earthquakes, scarps or cracks.

13.5.3   *The asthenosphere*

The asthenosphere is the highly viscous, mechanically weak and ductilely deforming region of the mantle which accommodates strain through plastic deformation. It is involved in plate tectonic movement and isostatic adjustments (Isostatic movement is considered in section 13.5.5). It is the (often localised) melting in the mantle which gives rise to magma,[750] which may then rise through the lithosphere and crust[751,752] and be extruded, as lava, onto the surface of the earth.[753]

The asthenosphere's average temperature is close to the melting point of some forms of rock, in particular olivine, but there is no absolute boundary between the lithosphere and asthenosphere, the one 'blends' into the other. The asthenosphere is, generally, found at depths of between approximately 80 and 200 km below the surface. Its thickness depends mainly on its temperature but it can, in some regions, extend as deep as 700 km (430 mi).[754]

The asthenosphere moves and the extent of its movement and deformation can be measured in centimetres per year over lineal distances. Overall, these movements may, eventually, measure thousands of kilometres.[755]

13.5.4   *Magma, or mantle, plumes*

Magma is a mixture of molten or semi-molten rock, volatiles and solids that is found beneath the surface of the Earth. Whether or not rock melts to form magma is controlled by three physical parameters; its temperature, pressure, and composition.[756] Magma rises toward the Earth's surface when it is less dense than the surrounding rock and when a structural zone, such as a fault, allows for movement. It often collects in magma chambers which may feed a volcano, or turn into a pluton.[757,758]

Mantle plumes, also referred to as magma plumes,[759] are columns of abnormally hot, semi-solid material that originate deep in the mantle, probably at the core–mantle boundary[760] and which rise through the Earth's mantle becoming a diapir[761] in the Earth's crust.[762] They may partially melt at shallow depths. Figures 13.2 and 13.3 give some idea of the concept of a mantle plume.

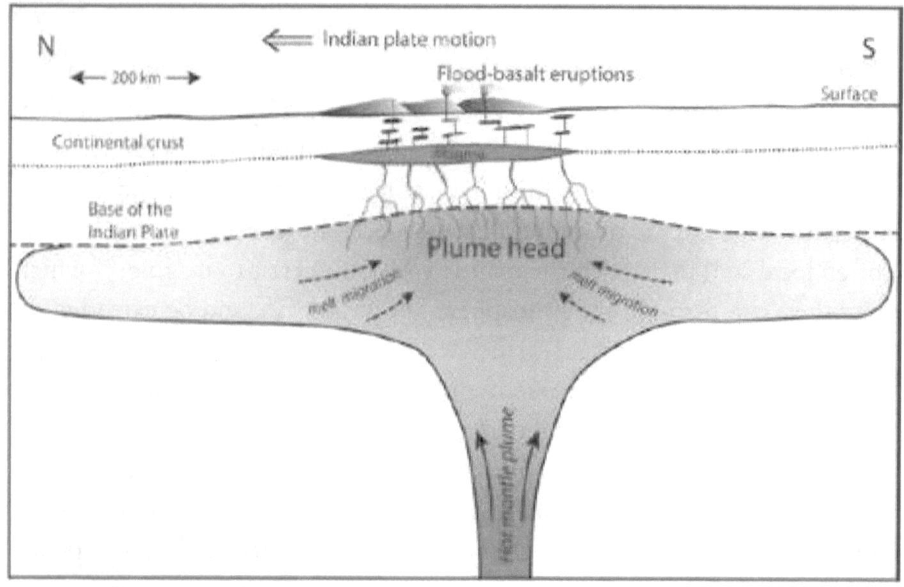

Figure 13.2 A hypothesised mantle plume under India. [763]

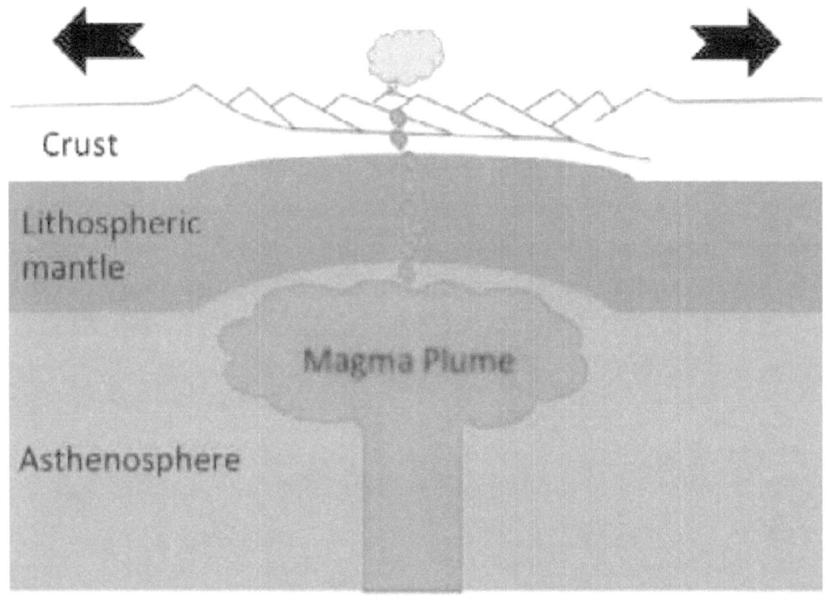

Figure 13.3 Mantle plume causing continental rifting. [764]

13.5.5 *Isostasy and isostatic adjustment*

Wikipedia defines isostasy as being *"the state of gravitational equilibrium between Earth's crust and mantle such that the crust 'floats' at an elevation that depends on its thickness and density"*.[765]

Due to the fact that the asthenosphere is not solid (as discussed previously), when any large mass (such as sediment or ice caps) is deposited on a particular region, the weight of it may cause the crust below to deform or sink. Figure 13.4 gives an idea of what happens when a large weight is placed on the surface of the earth. This is known as isostatic depression.[766] The reverse is also true so that when large amounts of material are removed or eroded away from a region the land may rise, or rebound, to compensate. The all-embracing term for this movement, whether up or down, is isostatic adjustment.

Isostatic post-glacial rebound[767] is what happens when ice-sheets melt, the load on the lithosphere and asthenosphere is reduced and, as a consequence, the surface and underlying layers rebound, attempting to reach equilibrium levels. The rebound movements are so slow that the uplift and subsidence caused by the ending of the last glacial period about

12 000 years ago is still continuing.[768] This is, I emphasise, the approximate time period since the stated date of the destruction of Atlantis.

Isostatic adjustment, whether rebound or subsidence, can also (indirectly and as a result of rising or dropping land and seabed levels) cause rising or dropping sea levels both locally and globally. By way of example, when the sea floor rises (as it is continuing to do in parts of the northern hemisphere) water is displaced and so has to go elsewhere.

It, isostatic adjustment, may also involve horizontal movements and can cause changes in Earth's gravitational field and rotation rate, polar wander, and earthquakes.[769]

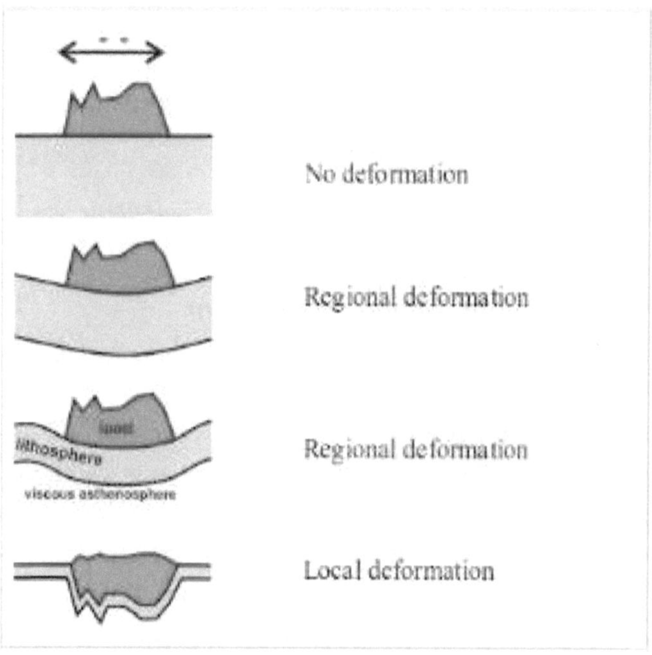

Figure 13.4 Diagram showing the isostatic vertical motions of the lithosphere (grey) in response to a vertical load (in green).[770]

13.5.6 Eustasy

Another relevant concept, considered (but not mentioned by name) in the previous section, is that of eustasy. According to Wikipedia eustasy *"refers to changes in the amount of water in the oceans, usually due to global climate change.*"[771] The level of the oceans can be affected by both a warming and a cooling climate.

When Earth's mean temperature increases there is, in addition to the melting of the ice with a concomitant increase in sea level,[772] a thermal expansion of sea water which also contributes to raising the level. Since the Last Glacial Maximum (about 22 000 years ago) the mean sea level has risen, as a result of the warming of Earth's temperature, by more than 120 m. This is primarily as a result of the melting of major ice sheets.[773,774]

Similarly, when Earth's climate cools a greater proportion of water is stored on the land masses in the form of glaciers, snow, etc. and this results in falling global sea levels (relative to a stable land mass).

13.6 Possible Mechanisms of submersion

Having given these brief explanations of these various concepts I shall now move on to consider them as possible mechanisms contributing to the submersion of Hatton Rockall. I advance them on the basis that I perceive them to be logically linked and in the hope that others, more suitably qualified, will take it further.

13.6.1 *Eustasy as a mechanism*

The one, immediate and undisputed, contributing factor to the submersion of Hatton Rockall would have been the eustatic sea level rise caused by the melting of the ice caps and flooding. Given, however, that the sea level has risen by only just about 120 metres since the last glacial maximum and that Hatton Rockall has, ostensibly, sunk by about 1 500 metres, this cannot have been the primary mechanism of its submersion.

13.6.2 *Isostatic adjustment as a mechanism*

The most substantial cause of the continued subsidence of Hatton Rockall was, I propose, isostatic adjustment. This, as set out previously, is generally caused by the removal, or unloading, of a weight on the surface of the earth. What is important for my proposition is where, how and when this adjustment has taken place.

13.6.2.1 Has there been isostatic adjustment in the north Atlantic region?

It is uncontentious that there were massive ice caps covering northern North America and northern Europe at the time of the glacial maximum. It also appears to be accepted that at the beginning of the glacial period and as the load of the ice caps on both North America and Europe increased, molten mantle material would have flowed, at depth, away from the areas covered by the ice caps and towards the periphery. The periphery would have included the North Atlantic area.[775] Conversely, it is also accepted that as the ice caps melted at the end of the Pleistocene (and sea levels increased) the resulting changed stresses[776] would have led to

> ➤ a mantle flow away from the outer areas and towards the centre of the area (from which the ice had melted and disappeared) and

> ➤ subsidence in the peripheral areas and corresponding uplift, or rebound, in the area formerly covered by the ice.[777]

Blundell and Waltham, in their article *A possible glacially-forced mechanism for late Neogene surface uplift and subsidence around the North Atlantic*[778] explain that, with deglaciation, there is a mantle flow away from the periphery and towards the area which had formerly been glaciated which results in a thickening of the crust in the formerly glaciated areas and a thinning of the crust on the periphery. This, in turn, creates subsidence on the periphery and rising land in the area formerly under the ice. It follows, as a matter of logic, that if there is subsidence on the periphery then that periphery must have been higher, or raised, before it started subsiding.

What is also of particular significance for my proposal is that, according to the authors, the time period for such a process to be completed, and for isostasy to be reached, is approximately 12 000 years - it is now almost 12 000 years since Atlantis is said to have been flooded. As is apparent from Figure 13.5, West Ireland and the south west and south of England are still showing relative subsidence[779] but the rate of subsidence is quite slow.

13.6.2.2   Was Hatton Rockall on the periphery of an ice sheet?

At the last glacial maximum vast amounts of the northern hemisphere were covered by icecaps. The Fennoscandian ice sheet extended, at one stage, from eastern Russia as far as Britain and Ireland and the Laurentide ice sheet covered most of Canada and a large portion of the northern United States. As far as we know the Hatton Rockall area, on the western edge of the Fennoscandian ice sheet, remained ice free.

As the Fennoscandian ice sheet melted, individual areas were left under ice, one such remaining area was over Britain and Ireland. According to the authors of the article *Pattern and timing of retreat of the last British-Irish Ice Sheet*[780] the British-Irish Ice Sheet (BIIS) attained its maximum extent by about 25 000 BCE and but separated from the Norwegian ice sheet prior to 23 000 BCE. Over time the BIIS separated into residual Irish and British ice sheets and ultimately melted away. The main focus of the weight of the BIIS (before it melted away) would have been more or less concurrent with the rising area shown in Figure 13.5. (To give an idea of the size and weight of the BIIS, its volume was, at its maximum size, sufficient to have raised global sea levels by about 2.5 metres, or 8 feet.[781])

Hatton Rockall was on the periphery of the western extreme of the Fennoscandian ice sheet and later, on the periphery of the British Irish ice sheet. It therefore falls within the purview of the findings of Blundell and Waltham's study, mentioned in the previous subsection. This would, concomitantly, suggest that Hatton Rockall would have been significantly higher in the water at the end of the last glacial maximum and probably above water until the flooding and the isostatic subsidence (consequent upon the isostatic uplift on the British Isles and the European continent) took place.

It could, as such, have been Plato's island of Atlantis.

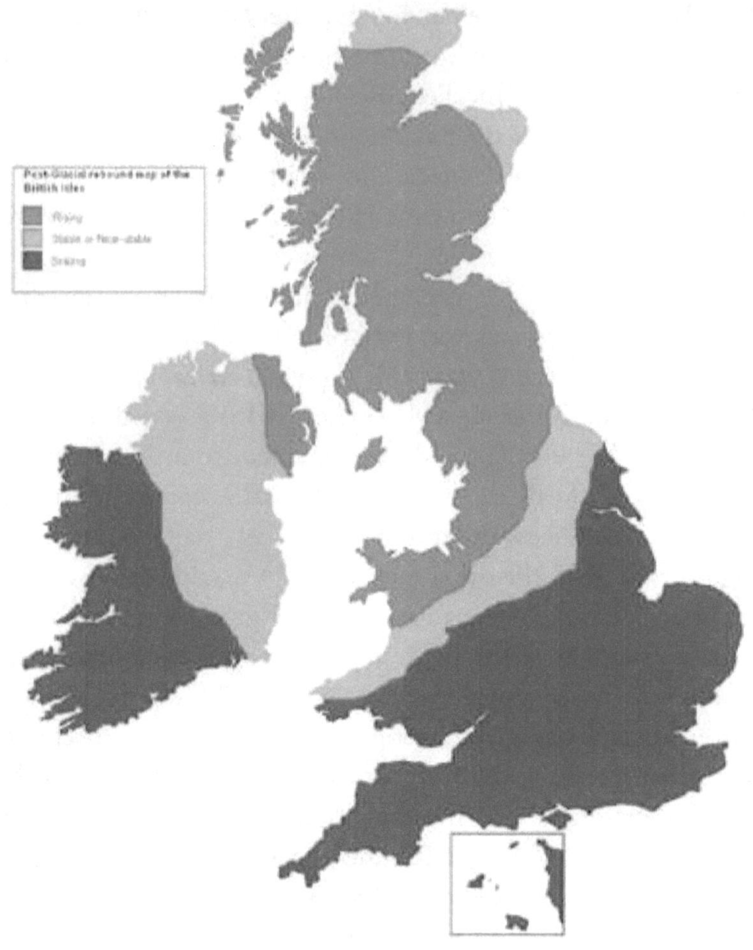

Figure 13.5 Post-glacial isostatic adjustment in the British Isles. [782]

13.6.3   *Reduction of mantle plume / magma as a mechanism*

In this section I shall consider whether:

1. there is any precedent for a large area of land being raised by the underlying asthenosphere and/or a mantle plume;[783]

2. there are factors from which one can draw the inference that there was a mantle plume, or magma body, of sufficiently large size, under Hatton Rockall and/or there is any indication that there had been volcanism in the Hatton Rockall area;

3. there can be a link between bodies of magma, or other causal factor, which could lead to the reduction of the volume of a supporting mantle plume,

4. there is any indication that the emptying of (or reduction in the volume of) a magma body could lead to the subsidence of land?

13.6.3.1 Is there any precedent for a large area of land being supported by underlying magma?

In a 2014 study[784] of the Atlas Mountains in Morocco which appeared in the journal *Geology*, Miller and Becker concluded that the Atlas Mountains of Morocco are anomalous in that they have an unusually high topography with only modest tectonic shortening and, further (and most importantly for my proposition), that the current topography of the Atlas mountain region is supported by upper mantle convection and upward mantle flow possibly affected by and related to the Canary hotspot.[785, 786] In order to have some understanding of the implications of this one has to have some idea of the extent of the range. The Atlas Mountain range, not all of which falls within Morocco, extends across the north-western stretch of Africa for about 2 500 km.[787] It traverses Morocco, Algeria, and Tunisia and is divided into a number of 'sub-ranges'. The High Atlas in Morocco, which includes the high point of Jbel Toubkal,[788] rises in the west at the Atlantic Ocean and stretches in an easterly direction to the Moroccan-Algerian border.[789] It has a number of peaks reaching above 4 000 m.[790] The Middle Atlas is the northernmost and second highest of three main Atlas Mountains chains of Morocco. It extends for about 350 km in a north-easterly direction.[791]

It has also been proposed that the Scandinavian mountains, the Scandes, are, or have been, supported by a body of magma.[792,793] Also, and although not entirely apposite here, a 2014 study[794] found that massive amounts of magma, which may ultimately give rise to a super volcano, may be stored in a horizontally layered reservoir.

If whole ranges of mountains may be supported by a body of magma then it appears to me to be possible that a large island may also be supported by such a body

13.6.3.2   Was there a magma chamber / plume, of sufficiently large size, under Hatton Rockall?

Firstly it should be reiterated that Hatton Rockall is situated in a volcanic passive margin and, secondly there are a number of structures in the area which have volcanic origins. Rockall Islet itself is described in Wikipedia as making up the "*eroded core of an extinct volcano (a volcanic plug), and is one of the few pinnacles of the surrounding Helen's Reef*".[795] Hasselwood Rock, which is less than a kilometre from Rockall Islet is described by Wikipedia as being "*the destroyed cone of an extinct volcano.*"[796] That said, it must be recognised that simply because there are volcanic structures present does not, of itself, provide sufficient grounds to think that the drainage of any magma cavities would be sufficient to cause the submersion of what must have been a very large island. For this one needs to look further to a much larger structure such as the Iceland plume.

"*The Iceland plume is a postulated upwelling of anomalously hot rock in the Earth's mantle beneath Iceland. Its origin is thought to lie deep in the mantle, perhaps at the boundary between the core and the mantle at ca. 2880 km depth.*"[797] It would appear that, although there is some disagreement between scholars as to the precise mechanism of the magma body and how deep it extends, there is agreement that a very large body of magma exists beneath Iceland and the surrounding area. According to White and McKenzie,[798] the width of the magma body under Iceland is approximately 2 000 km.

The body is not, it appears, simply one 'blob'. Hartley, Ross A., *et al.*,[799] writing in *Nature Geoscience*, hypothesise that there are 'plume conduits' up which short-period (approximately million years) thermal anomalies spread out as solitary waves, from the plume centre, for distances of approximately 1 000 km.[800,801] Such fluctuations in the temperature of the plume can, according to Poore *et al.*,[802] change surface elevations by tens to hundreds of metres and that, at a distance of 600 kilometres from the centre, an increase of temperature by 25⁰C can cause an uplift of between 90 to 180 metres.[803] In to a 2005 study by Arrowsmith *et al.*[804] the authors comment the Iceland plume "*has had an important influence on vertical motions in the North Atlantic.*"

Given this it and other articles[805] it appears to be uncontentious that a plume, or magma body, gives support and can cause uplift.

Arrowsmith *et al.* also found that an arm of plume material extends out from the Iceland plume and beneath the British Isles. In 2012, Davis *et al.*[806] published a study in which they considered the findings of Arrowsmith *et al.* and concluded that Arrowsmith *et al's* findings were consistent with magmatic underplating and that a 'warm' offshoot of asthenosphere from the Iceland plume could provide sub-lithospheric isostatic support.[807] Similarly Rickers *et al.*[808] concluded that their model provides strong support for the concept that there is wide-spread dynamic support of the North Atlantic region by the Iceland plume. They also note that there are two 'fingers' extending from the hotspot, one of which extends along the British Isles to Brittany. This finger would, therefore, pass under, or close to, the Hatton Rockall area.

Pfeffer, *et al*[809] recognise (albeit that they are considering a lot longer timeframes than I am considering here) that it appears that mantle flow and structure may play a dominant role in surface motions.[810]

It is therefore apparent that a large body of magma runs beneath, or close to, Hatton Rockall and that such a large body of magma may affect the level of the earth's surface.

13.6.3.3   Can bodies of magma be linked?

Any links (if Hatton Rockall itself was supported by magma) would have to have been large and have extended over a much larger area.

Petford *et al.*,[811] writing in the journal *Nature,* reviewed the physical processes controlling how granitic melts are extracted from their source region, transported through and intruded into pre-existing rock and concluded that they favour the concept of the transport of magma through narrow conduits which may take the form of dykes, pre-existing faults or shear zones. According to them the ascent of magma through dykes can be a million times faster than would be expected in terms of the more traditional ideas of a slow rising diapir of magma. They find that models operating on time scales of months to years are replacing older models requiring millions of years for completion. They also comment that recognition is now being given to the important role played by tectonic activity in making spaces for magma, and, further, that the emplacement of magma is episodic and involves separate pulses of magma. Similarly Cembrano, J *et al.*[812] talk of magma plumbing systems. They conclude, amongst other things, that crustal thickness is an important factor in the link between tectonics and volcanism.[813]

Petford *et al.*'s comment that the, presumably large scale, movement of magma may take place on time scales of month to years (rather than millions of years) also lends credence to the possibility that magmatic support of Hatton Rockall could have moved away in a matter of thousands, rather than millions, of years.

Whilst not directly related to the North Atlantic, Duggen *et al.*,[814] propose the possible existence of a subcontinental link between the Canary island mantle plume and the Mediterranean. They also, importantly for my proposition, propose that it is possible for magma, forming part of the mantle, to be subjected to suction and to flow backwards as a result, thus depleting some regions of sub-crustal magma. This 'sucking-back' effect, particularly in the context of isostatic rise of an area previously under the ice cap[815], would probably compound the subsidence of the peripheral area.

Although not specifically dealing with the interlinking of magma bodies a study by Chaussard and Amelung,[816] found that differences in the density and structure of the crust may influence levels of magma storage. In their discussion the authors conclude that it is only upper plate crustal settings, which would include the thickness and density of the upper crust, which control magma ascent and storage. It follows, logically, from Chaussard and Amelung's finding that if the density, thickness or weight of the crust were to change then the rate of ascent of magma would be changed. If the weight of the crust was increased (by, for example, the increased weight of flood waters) the rate of the rise of magma would, logically, decrease or even, I suggest, become negative (i.e. the magma would descend).

13.6.3.4 Can the movement or abstraction of magma lead to the subsidence of land?

Larsen and Sønderholm[817] found that parts of Greenland underwent rapid, major subsidence (during the Palaeocene) which was associated with extensive volcanism. Similarly Boldreel and Andersen,[818] found that eruptions which deplete subsurface magma reservoirs can produce local subsidence (albeit that their study also concerned an earlier era).

If this could have happened then, then it appears to me to be logical that, it could also have happened early in the Holocene, particularly as there was, as discussed earlier, extensive volcanism around the time of the submersion of Atlantis. .

Although on a much smaller scale, individual volcanoes are also known to experience cycles of slow uplift followed by sudden subsidence. This has been recorded at various volcanoes including Kilauea, Mauna Loa, and Krafla. It appears that that magma has repeatedly filled and then withdrawn from a reservoir beneath the volcano.[819, 820]

It then becomes plausible, given the above, that the removal of, or reduction in, supporting magma and/or molten mantle material could have been a contributing factor in the ongoing subsidence of Hatton Rockall, or Atlantis, after the flooding.

13.6.4 *The weight of the flood waters as a contributing mechanism.*

Given the findings by Blundell and Waltham, the map supplied by NOAA showing Hatton Rockall at 16 000 BCE (See section 7.2), the unconformities in the veneer of sediment over Hatton Rockall, the gypsum found and the hiatuses in the growth of cold water corals during the ice ages, it appears difficult for it to be disputed that Hatton Rockall has been subaerial in the past and, in particular, that it probably was subaerial at the end of the last ice age.

The problem that stands in the way of this being readily accepted must, it appears to me, be the rate at which a subaerial Hatton Rockall would have had to have subsided to reach the depths, at which it now is, within ~12 000 years. Although there are precedents (as we have seen in section 13.3) for land moving at an unusually fast rate, there must still be some underlying reason behind the fast movement. A reason, in this case, might be the effect of the added weight of the melted ice and flood waters acting in conjunction with the all the various factors discussed.

With deglaciation, the oceans would have filled causing an increased water load on the ocean floor whilst the continental regions would have lost their load and would have rebounded, or become uplifted, as discussed above. The study by Chaussard and Amelung,[821] (mentioned in section 13.6.3.3) found that differences in, in effect, the weight and structure of the crust influences levels of magma storage and magma ascent and descent. Changes in magma ascent and descent could, in turn, have affected the level of the surface of the crust - and possibly led to its subsidence. Given the vast volumes of water added to the North Atlantic it is probable that their weight would have affected the rate at which magma ascended or descended. This would, concomitantly and as we have seen, have affected the level of the land (or sea-bottom) surface.[822] Frederikse *et al*[823] have found that various areas of the ocean bottom are still deforming due to mass distribution – which includes the result of water added by presently melting glaciers.

Shaw, Champion *et al.*,[824] in another study about the nearby Faeroe Shetland basin, concluded that a water load may cause the subsidence of the oceanic crust and, importantly, that the rates of *"water-loaded tectonic subsidence following peak uplift are several times greater"* than the normally expected maximum subsidence rate. The accelerated rate of subsidence is, of itself, important but they also propose that the best explanation for this phenomenon of accelerated subsidence is that it is caused by *"a mantle convective phenomenon"*. They propose that a region of hot plume material, which had produced transient uplift (in an earlier time), was advected away resulting in the subsidence (or increased rate of subsidence) of the relevant area.

It is, therefore, apparent that the added weight from the melted ice and of the flood waters would have accelerated the rate of subsidence and made it faster than might normally be expected.

13.7 Conclusion

In conclusion it is apparent that
    (a) there was a rebound of the areas which had been under an ice sheet with a concomitant sinking of the peripheral areas and

        i. Hatton Rockall lies on the periphery of, at least, the area covered by the British Irish Ice Sheet,

ii. Hatton Rockall, as part of the peripheral area would (along with parts of the western coast of Ireland) and as a result of the isostatic adjustment caused by the melting of the ice, have naturally subsided

(b) there was massive volcanism at the time which could have led to a reduction of the Iceland Plume and some, concomitant, reduction in support of the Hatton Rockall area,

(c) with the rebound, of the areas formerly under the ice sheet, there would have been a flow of the viscous mantle away from the periphery thereby causing a thinning of the crust and, additionally, adding to the subsidence,

(d) the weight of the water covering Hatton Rockall would have increased the rate of subsidence,

(e) there are precedents for land sinking, and moving laterally, at speeds similar to (and in excess of) what would be required in order for Hatton Rockall to have sunk within 10 000 to 12 000 years and

(f) it has been found that it would take approximately 12 000 (which we are now reaching from the time that Plato said Atlantis sunk) years for an area to reach equilibrium after removal of ice sheets.

Given all these factors I propose that it is possible, and likely, that the sub-aerial Hatton Rockall area could

➤ have subsided at a rate unknown in recorded history and, consequently,

➤ have reached its present depths within a period of about ten to twelve thousand years.

SECTION FOURTEEN:
APPENDIX:
DEPTH PROFILES OF HATTON ROCKALL

Below are a number of pictures[825] with inserted graphs showing the profile of Hatton Rockall along the marked line. As can be seen from the profile graphs the drop-off on the 'seaward' sides is generally steep and precipitous.

The first two images are included simply for reference purposes and show Hatton Rockall itself and then it in its geographical context. They do not include any profile graph.

In all of the pictures, save A.2, North is at approximately the top of the picture. In A.2 north is to the left of the page.

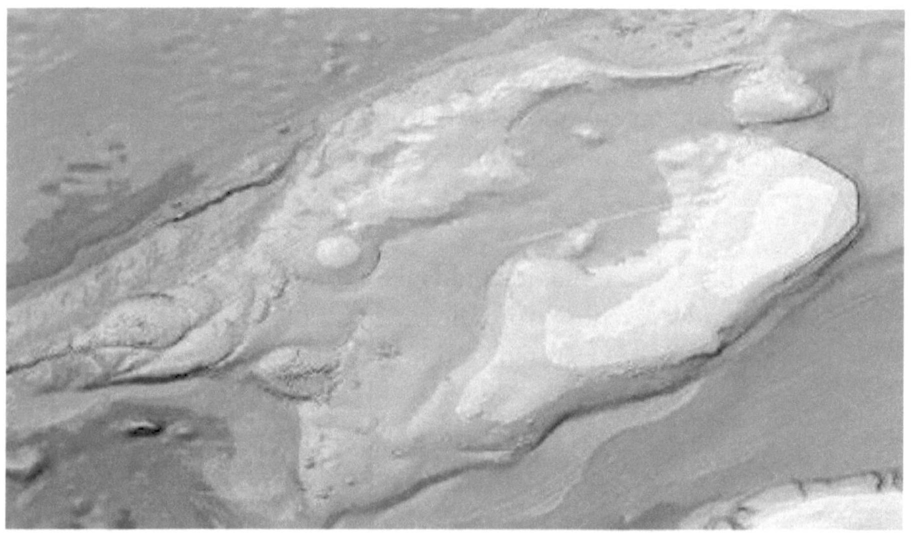

Figure A.1 Hatton Rockall

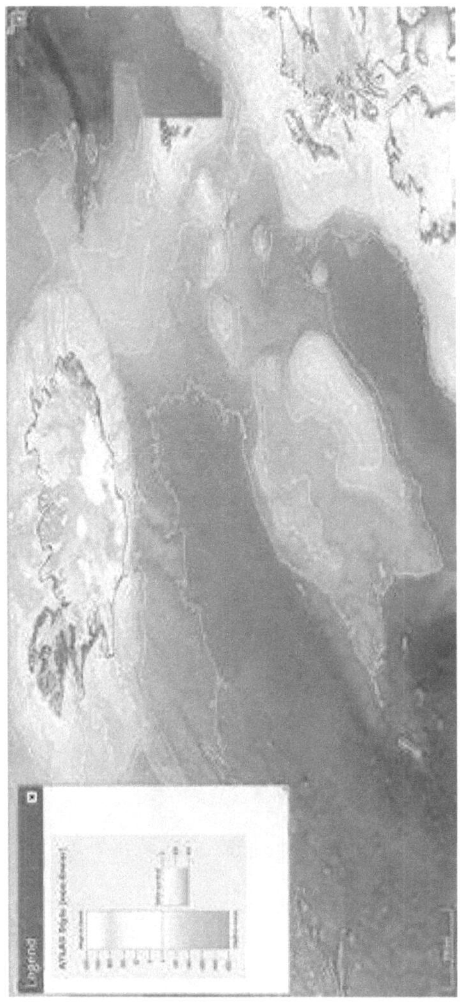

Figure A.2 Hatton Rockall in its geographical and bathymetric context

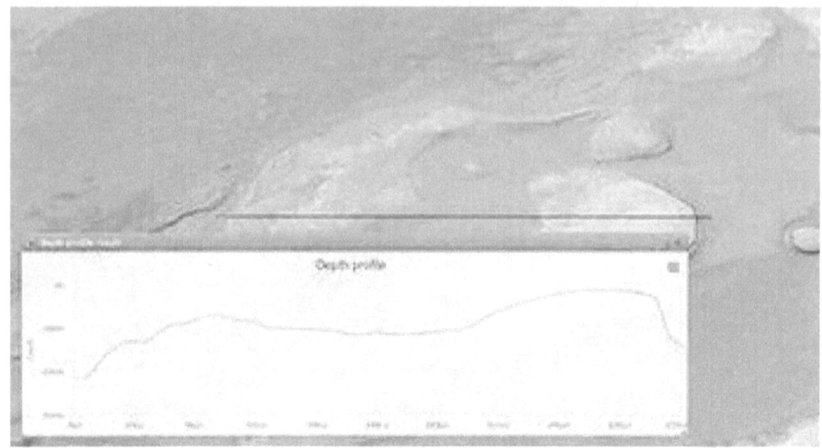

Figure A.3 West to East profile across Hatton Bank, the Hatton Rockall Basin and Rockall Bank

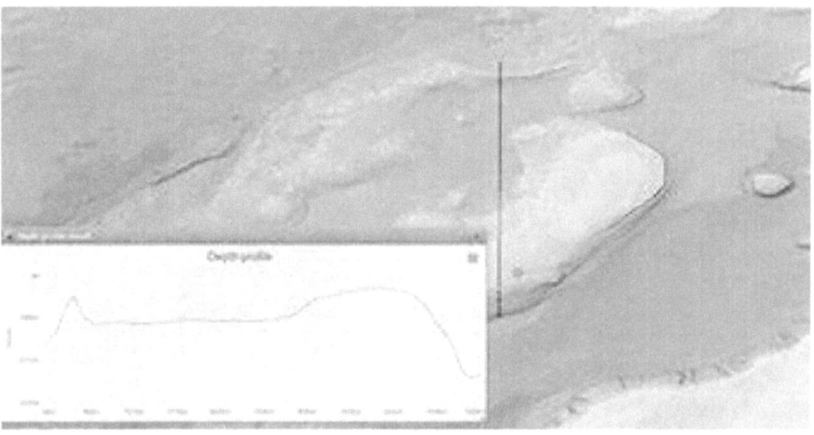

Figure A.4 North to South across Hatton Rockall Basin showing steep slope on 'seaward' sides (both northern and southern) and the high mountains to the north of the 'plain'.

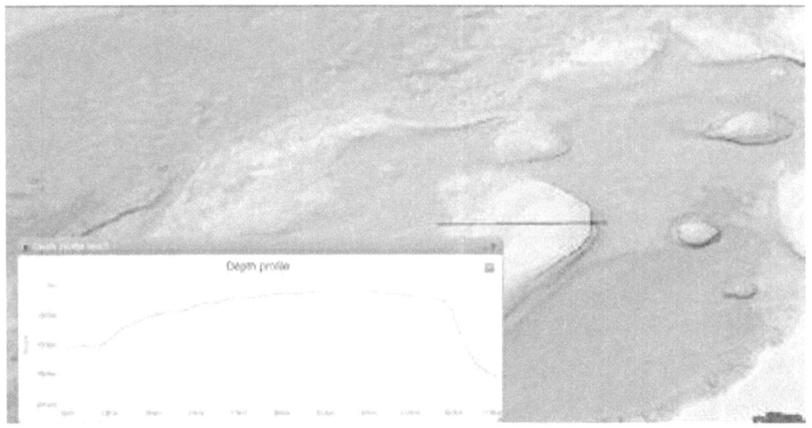

Figure A.5 West to East profile across Rockall Bank showing steep slope on 'seaward' (eastern) side.

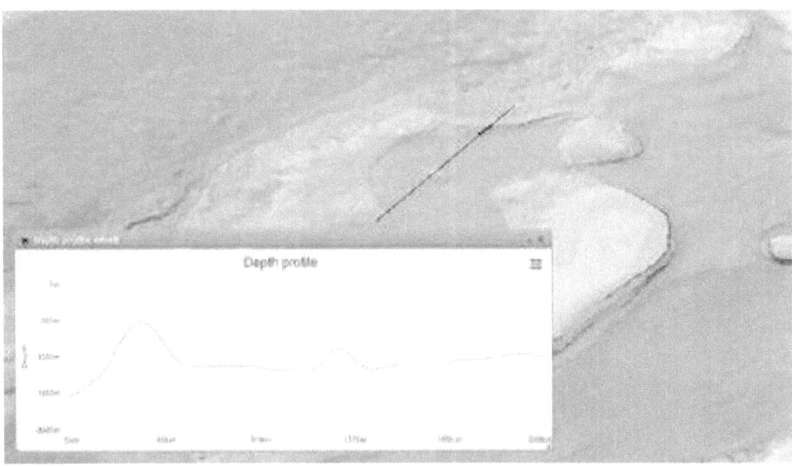

Figure A 6 North East to South West profile across Hatton Rockall Basin and through central Mammal Mount.

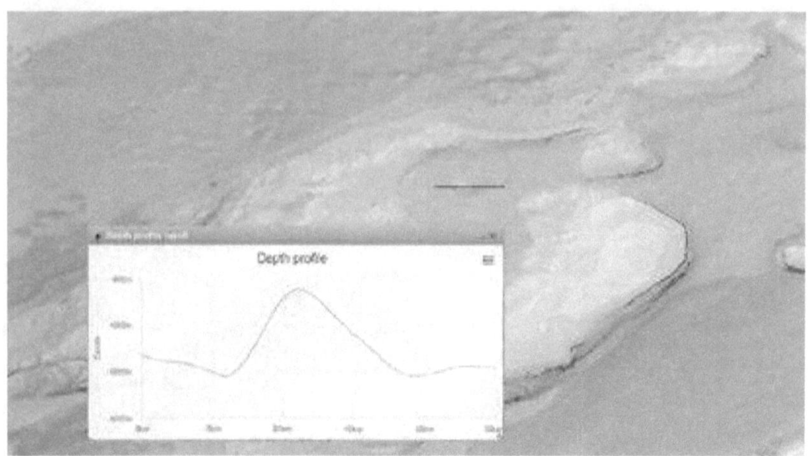

Figure A 7 West to East profile across Mammal Mount showing dipping/moating at its sides

Bibliographic References
(Books and articles)

Achilli, A. et al. *"Saami and Berbers – an unexpected mitochondrial DNA link."* The American Journal of Human Genetics 76.5, 2005, 883-886

Adams, J. *"Postglacial faulting in eastern Canada: nature, origin and seismic hazard implications."*, Tectonophysics 163.3, 1989, 323-331

Aitchison, J. C., J. R. Ali, and A. M. Davis. *"When and where did India and Asia collide?"* Journal of Geophysical Research: Solid Earth (1978–2012) 112.B5, 2007

Alley, R. B. *"The Younger Dryas cold interval as viewed from central Greenland."* Quaternary Science Reviews 19.1, 2000, 213-226

Alley, R. B. *"Ice-core evidence of abrupt climate changes."* Proceedings of the National Academy of Sciences 97.4, 2000, 1331-1334

Alonso, S. et al. *"The place of the Basques in the European Y-chromosome diversity landscape."* European journal of human genetics 13.12, 2005, 1293-1302

Alonso-Garcia, M., F. J. Sierro, and J. A. Flores. *"Arctic front shifts in the subpolar North Atlantic during the Mid-Pleistocene (800–400ka) and their implications for ocean circulation."* Palaeogeography, Palaeoclimatology, Palaeoecology 311.3, 2011, 268-280.

Anderson, D. L. and S. D. King. *"Driving the Earth machine?"* Science 346.6214, 2014, 1184-1185.

Appollodorus. (Ed. Sir James G. Frazer) *Book 2 of the Library by Appollodorus;* Harvard University Press, 1921

Arrowsmith, S. J. et al. *"Seismic imaging of a hot upwelling beneath the British Isles."* Geology 33.5, 2005, 345-348

Arslanov, Kh A. et al. *"Consensus dating of mammoth remains from Wrangel Island."* Radiocarbon 40.1, 1998, 289-294

Arvidsson, R. *"Fennoscandian earthquakes: whole crustal rupturing related to postglacial rebound."* Science 274.5288, 1996, 744-746

Babcock, W. H. *Legendary Islands of the Atlantic: A Study in Medieval Geography. No. 8.* American Geographical Society, 1922.

Backman, J. *"Miocene-Pliocene nannofossils and sedimentation rates in the Hatton-Rockall basin, NE Atlantic Ocean."* An inaugural dissertation, 1980

Bacon, S. N., and S. K. Pezzopane. *"A 25,000-year record of earthquakes on the Owens Valley fault near Lone Pine, California: Implications for recurrence intervals, slip rates, and segmentation models."* Geological Society of America Bulletin 119.7-8, 2007, 823-847

Baeckstroem, A., N. Rantakokko, and M. V. Ask. *"Structural Analysis of the Pärvie Fault in Northern Scandinavia."*. AGU Fall Meeting Abstracts. Vol. 1.1, 2011

Baldi, P. and B. R. Page. *"Europa Vasconica-Europa Semitica: Theo Vennemann, Gen. Nierfeld, in: Patrizia Noel Aziz Hanna (Ed.), Trends in Linguistics, Studies and Monographs 138, Mouton de Gruyter, Berlin, 2003, pp. Xxii+ 977."* Lingua 116.12, 2006, 2183-2220

Barber, E. W. and P. T. Barber. *When they severed earth from sky: how the human mind shapes myth.* Princeton University Press, 2006

Behar, D. M., et al. *"The Basque paradigm: genetic evidence of a maternal continuity in the Franco-Cantabrian region since pre-Neolithic times."* The American Journal of Human Genetics 90.3, 2012, 486-493

Belka, Z. *"Early Devonian Kess-Kess carbonate mud mounds of the eastern Anti-Atlas (Morocco), and their relation to submarine hydrothermal venting."* Journal of Sedimentary Research 68.3, 1998

Berndt, C. et al. *"Kilometre-scale polygonal seabed depressions in the Hatton Basin, NE Atlantic Ocean: Constraints on the origin of polygonal faulting."* Marine Geology 332, 2012, 126-133

Björck, S. et al. *"The Preboreal oscillation around the Nordic Seas: terrestrial and lacustrine responses."* Journal of Quaternary Science 12.6, 1997, 455-465

Blundell, D. J. and D. A. Waltham. *"A possible glacially-forced tectonic mechanism for late Neogene surface uplift and subsidence around the North Atlantic."* Proceedings of the Geologists' Association 120.2, 2009, 98-107

Boldreel, L. O. and M. S. Andersen. *"Tertiary development of the Faeroe-Rockall Plateau based on reflection seismic data."* Bulletin of the Geological Society of Denmark 41.2, 1994, 162-180.

Bonow, J. M., et al. *"Elevated erosion surfaces in central West Greenland and southern Norway: their significance in integrated studies of passive margin development."* Norwegian Journal of Geology 87, 2007, 197-206

Boutkan, D. and M. G. Kossmann. *"Some Berber parallels of European substratum words."* Journal of Indo-European Studies 27.1/2, 1999, 87

Bowler, Jim M., et al. *"Pleistocene human remains from Australia: a living site and human cremation from Lake Mungo, western New South Wales."* World archaeology 2.1 (1970): 39-60.

Brandes, C., U. Polom, and J. Winsemann. *"Reactivation of basement faults: interplay of ice-sheet advance, glacial lake formation and sediment loading."* Basin Research 23.1, 2011, 53-64

CELT: Corpus of Electronic Texts, *Briefe relation of Ireland, and the diversity of Irish in the same [and] Priests in Ireland and Gentlemen gone abroad'.* http://www.ucc.ie/celt/published/T100077.html 2014

Briggs, R. W., and S. G. Wesnousky. *"Late Pleistocene and Holocene paleoearthquake activity of the Olinghouse fault zone, Nevada."* Bulletin of the Seismological Society of America 95.4, 2005, 1301-1313

Brothers, D. S., K. M. Luttrell, and J. D. Chaytor. *"Sea-level–induced seismicity and submarine landslide occurrence."* Geology 41.9, 2013, 979-982

Bryant, E., G. Walsh, and D. Abbott. *"Cosmogenic mega-tsunami in the Australia region: are they supported by Aboriginal and Maori legends?"* Geological Society, London, Special Publications 273.1, 2007, 203-214

Buckley, T. A. *The Odyssey of Homer.* Burt, 1899

Budge, E. A. T. W. *The Gods of the Egyptians, Or Studies in Egyptian Mythology, by EA Wallis Budge* Methuen, 1904

Burges, G "The Works of Plato, a new and literal version. Vol. IV containing the doubtful works, Lives of Plato, by Diogenes Laertius, Hesychius and Olympiodorus

; Introductions to his doctrines, by Alcinous and Albinus. Published by H.G. Bohn, (1854).

Bury, R.G. (Translator) *Plato. Letter No. 7 from "Plato. Plato in Twelve Volumes", Vol. 7*, Cambridge, MA, Harvard University Press; London, William Heinemann Ltd., 1966.

Bury, R.G. (Translator) *Plato, Phaedrus from "Plato. Plato in Twelve Volumes", Vol. 7*, Cambridge, MA, Harvard University Press; London, William Heinemann Ltd., 1966.

Butcher, SH & Lang, S (Translators) *The Odyssey of Homer.* Macmillan, 1895

Butler, S. *The Odyssey render into English prose for the use of those who cannot read the original.* 1922.

Campion, S. E. *A Historie of Ireland, written in the Yeare 1571.* Hibernia Press, 1809

Carminati, E. and C. Doglioni. *"North Atlantic geoid high, volcanism and glaciations." Geophysical Research Letters 37.3,* 2010

Cashman, Katharine V., and Shane J. Cronin. *"Welcoming a monster to the world: Myths, oral tradition, and modern societal response to volcanic disasters." Journal of Volcanology and Geothermal Research 176.3,* 2008, 407-418

Cembrano, J. and L. Lara. *"The link between volcanism and tectonics in the southern volcanic zone of the Chilean Andes: A review." Tectonophysics 471.1.,* 2009, 96-113

Ceramicola, S., et al. *"Anomalous Cenozoic subsidence along the 'passive' continental margin from Ireland to mid-Norway." Marine and petroleum geology 22.9,* 2005, 1045-1067

Champion, M. E., et al. *"Quantifying transient mantle convective uplift: An example from the Faroe Shetland basin." Tectonics 27.1,* 2008

Chaussard, E. and F. Amelung. *"Regional controls on magma ascent and storage in volcanic arcs." Geochemistry, Geophysics, Geosystems 15.4,* 2014, 1407-1418

Chen, Y. *"Did the reservoir impoundment trigger the Wenchuan earthquake?." Science in China Series D: Earth Sciences 52.4,* 2009, 431-433

Clark, C. D., et al. *"Pattern and timing of retreat of the last British-Irish Ice Sheet." Quaternary Science Reviews 44,* 2012, 112-146

Clarkson, Chris, et al. *"Human occupation of northern Australia by 65,000 years ago." Nature* 547.7663 (2017): 306

Claudius Ælianus his various history. For Thomas Dring translated by T Stanley (1665)

Clift, P. D., and J. Turner. *"Dynamic support by the Iceland Plume and its effect on the subsidence of the northern Atlantic margins." Journal of the Geological Society 152.6,* 1995, 935-941

Collina-Girard, J. *"L'Atlantide devant le détroit de Gibraltar? Mythe et géologie." / "Atlantis off the Gibraltar Strait? Myth and Geology" Comptes Rendus de l'Académie des Sciences-Series IIA-Earth and Planetary Science 333.4,* 2001, 233-240

Comyn, E and Dineen, P. S. (Translators) *The History of Ireland by Geoffrey Keating (Foras Feasa ar Éireann le Seathrún Céitinn)*

Condron, A. and P. Winsor. *"Meltwater routing and the Younger Dryas." Proceedings of the National Academy of Sciences 109.49,* 2012, 19928 - 19933

Curry, Andrew, *"Seeking the Roots of Ritual"* Science 319(2008): 278-280

Cowper, W (Translator) *The Odyssey of Homer* London: J·M·Dent·&·Sons·Ltd and in New York by E·P·Dutton & Co, 1820

Curry, A. *"Seeking the Roots of Ritual," Science 319,* 2008, 278

Dai, M. et al. *"A Study on the Relationship between Water Levels and Seismic Activity in the Three Gorges Reservoir."* Institute of Seismology, China Earthquake Administration, Wuhan (tr. and pub in English by Probe International 2011), http://probeinternational. org/library/wp—content/uploads/ZOI1/06/3—Gorges—Report—26-5. Pdf, 2010

Dam, G., M. Larsen and M. Sønderholm. *"Sedimentary response to mantle plumes: implications from Paleocene onshore successions, West and East Greenland."* Geology 26.3, 1998, 207-210

Darwin, C. *"A naturalist's voyage round the world: Journal of researches into the natural history and geology of the countries visited during the voyage round the world of HMS Beagle under the command of Captain Fitz Roy."* RN, 1913

Davies, T. A., and A. S. Laughton. *"Sedimentary processes in the North Atlantic."* Initial reports of the deep sea drilling project 12, 1972, 905-934

Davis, M. W., et al. *"Crustal structure of the British Isles and its epeirogenic consequences."* Geophysical Journal International 190.2, 2012, 705-725

de Boer, J. Z., J. R. Hale, and J. Chanton. *"New evidence for the geological origins of the ancient Delphic oracle (Greece)."* Geology 29.8, 2001, 707-710.

De Camp, L. S. *Lost Continents.* Gnome Press, 1954

Dehls, J. F., et al. *"Neotectonic faulting in northern Norway; the Stuoragurra and Nordmannvikdalen postglacial faults."* Quaternary Science Reviews 19.14, 2000, 1447-1460

Delle Donne, D. et al. *"Earthquake-induced thermal anomalies at active volcanoes."* Geology 38.9, 2010, 771-774

Deur, Douglas. "A most sacred place: The significance of Crater Lake among the Indians of Southern Oregon." *Oregon Historical Quarterly* 103.1 (2002): 18-49.

Diefendorf, A. F. *Late-Glacial to Holocene climate variability in western Ireland. Diss. University of Saskatchewan,* 2005

Diodorus Siculus *The Library of History (Book V) of Diodorus Siculus, in Vol. III of the Loeb Classical Library edition,* Loeb Classical Library, 1939

Dixon, R.M.W., *"Origin legends and linguistic relationships"* Oceania 67, 1996, 127-139

Dlxon, R. M. W. *"The Dyirbal language of North Queensland."* Cambridge Studies in Linguistics, Cambridge University Press, 1972

Doré, A. G., et al. *"Principal tectonic events in the evolution of the northwest European Atlantic margin."* Geological Society, London, Petroleum Geology Conference series. Vol. 5., Geological Society of London, 1999, 41-61

Dorschel, B. et al. *"Growth and erosion of a cold-water coral covered carbonate mound in the Northeast Atlantic during the Late Pleistocene and Holocene."* Earth and Planetary Science Letters 233.1, 2005, 33-44

Duggen, S. et al. *"Flow of Canary mantle plume material through a subcontinental lithospheric corridor beneath Africa to the Mediterranean."* Geology 37.3, 2009, 283-286

Durá-Gómez, I. and P. Talwani. *"Reservoir-induced seismicity associated with the Itoiz Reservoir, Spain: a case study."* Geophysical Journal International 181.1, 2010, 343-356

Dvorak, J. J. and D. Dzurisin. *"Volcano geodesy: The search for magma reservoirs and the formation of eruptive vents."* Reviews of Geophysics 35.3, 997, 343-384

Echo-Hawk, Roger C. "Ancient history in the New World: integrating oral traditions and the archaeological record in deep time." *American Antiquity* 65.2 (2000): 267-290.

Edwards, C. J. et al. *"Dual origins of dairy cattle farming–evidence from a comprehensive survey of European Y-chromosomal variation." PLoS One 6.1*, 2011, e15922

Eggert, S. and T. R. Walter. *"Volcanic activity before and after large tectonic earthquakes: observations and statistical significance." Tectonophysics 471.1*, 2009, 14-26

Eisele, M. et al. *"Productivity controlled cold-water coral growth periods during the last glacial off Mauritania." Marine Geology 280.1*, 2011, 143-149

Elfstrom, A. *"The Båldakatj Boulder Delta, Lapland, Northern Sweden." Geografiska Annaler. Series A, Physical Geography, Vol. 65, No. 3/4*, 1983, 201-225

Elfström, Å. and L. Rossbacher *"Erosional Remnants in the Båldakatj Area, Lapland, Northern Sweden. A Terrestrial Analog for Martian Landforms" Geografiska Annaler. Series A, Physical Geography, Vol. 67, No. 3/4*, 1985, 167-176

Elias, S.A., Short, S.K., Nelson, C.H., Birks, H.H., *Life and times of the Bering land bridge. Nature 382, 60–63.* 1996

Ezat, M. M., T. L. Rasmussen, and J. Groeneveld. *"Persistent intermediate water warming during cold stadials in the southeastern Nordic seas during the past 65 ky." Geology 42.8*, 2014 , 663-666

Farr, T. G., C. Jones and Z. Liu. *"Progress Report: Subsidence in the Central Valley California "* Jet Propulsion Laboratory, California Institute of Technology, 2015

Fernández Domínguez, E. et al. *"Ancient DNA analysis of 8000 BC near eastern farmers supports an early neolithic pioneer maritime colonization of Mainland Europe through Cyprus and the Aegean Islands." PLoS Genetics, 2014, vol. 10, num. 6*, 2014, e1004401

Field, L., et al. *"Magma storage conditions beneath Dabbahu Volcano (Ethiopia) constrained by petrology, seismicity and satellite geodesy." Bulletin of volcanology 74.5*, 2012, 981-1004

Finlayson, B. et al. *"Architecture, sedentism, and social complexity at Pre-Pottery Neolithic A WF16, Southern Jordan." Proceedings of the National Academy of Sciences 108.20*, 2011, 8183 - 8188.

Firestone, R. *"The Case for the Younger Dryas Extraterrestrial Impact Event: Mammoth, Megafauna and Clovis Extinction." Journal of Cosmology 2*, 2009, 256-285

Fisher, T. G., D. G. Smith and J. T. Andrews. *"Preboreal oscillation caused by a glacial Lake Agassiz flood." Quaternary Science Reviews 21.8*, 2002, 873-878

Frank, N. et al. *"Deep-water corals of the northeastern Atlantic margin: carbonate mound evolution and upper intermediate water ventilation during the Holocene." Cold-water corals and ecosystems. Springer Berlin Heidelberg*, 2005, 113-133

Frank, R. M. *"Recovering European ritual bear hunts: a comparative study of Basque and Sardinian ursine carnival performances." INSULA: Quaderno di cultura sarda 3*, 2008, 41-97

Frazer, James George, ed. *Apollodorus: The Library*. Vol. 2. W. Heinemann, 1921.

Frederikse, Thomas, Riccardo EM Riva, and Matt A. King. *"Ocean Bottom Deformation Due To Present-Day Mass Redistribution and Its Impact on Sea Level Observations." Geophysical Research Letters 44.24* (2017).

Fregel, R., et al. *"The maternal aborigine colonization of La Palma (Canary Islands)." European Journal of Human Genetics 17.10*, 2009, 1314-1324

Fu, Qiaomei, et al. *"The genetic history of Ice Age Europe." Nature*, 2016

Geoffroy, L. *"Volcanic passive margins." Comptes Rendus Geoscience 337.16*, 2005, 1395-1408

Gibson, T. "Epilogue to Plato: The bias of literacy." *Proceedings of the Media Ecology Association. Vol. 6.,* 2005

Gonnermann, H. M., et al. *"Coupling at Mauna Loa and Kilauea by stress transfer in an asthenospheric melt layer." Nature Geoscience 5.11,* 2012, 826-829

González, A. M., et al. *"The mitochondrial lineage U8a reveals a Paleolithic settlement in the Basque country." BMC genomics 7.1,* 2006, 124

Green, P. F., et al. *"Thermochronology, erosion surfaces and missing section in West Greenland." Journal of the Geological Society 168.4,* 2011, 817-830

Grindon, A. J., and A. Davison. *"Irish Cepaea nemoralis land snails have a cryptic Franco-Iberian origin that is most easily explained by the movements of Mesolithic humans." PloS one 8.6,* 2013, e65792

Gupta, H. K. *"A review of recent studies of triggered earthquakes by artificial water reservoirs with special emphasis on earthquakes in Koyna, India." Earth-Science Reviews 58.3,* 2002, 279-310

Guthrie, R. D. *"Radiocarbon evidence of mid-Holocene mammoths stranded on an Alaskan Bering Sea island." Nature 429.6993,* 2004, 746-749.

Gutscher, M.-A. *"Destruction of Atlantis by a great earthquake and tsunami? A geological analysis of the Spartel Bank hypothesis." Geology 33.8,* 2005, 685-688

Haarmann, H. *"Indo-Europeanization—the seven dimensions in the study of a never ending process." Documenta Praehistorica XXXIV. Neolithic Studies 14,* 2007, 155-175

Hainzl, S., et al. *"Evidence for rainfall-triggered earthquake activity." Geophysical Research Letters 33.19,* 2006

Hampel, A., et al. *"Three-dimensional numerical modelling of slip rate variations on normal and thrust fault arrays during ice cap growth and melting." Journal of Geophysical Research: Solid Earth (1978–2012) 114.B8,* 2009

Hampel, A., R. Hetzel, and G. Maniatis. *"Response of faults to climate-driven changes in ice and water volumes on Earth's surface." Philosophical Transactions of the Royal Society A: Mathematical, Physical and Engineering Sciences 368.1919,* 2010, 2501-2517

Hansen, D. M. and J. Cartwright. *"The three-dimensional geometry and growth of forced folds above saucer-shaped igneous sills." Journal of Structural Geology 28.8,* 2006, 1520-1535

Hardiman, J. *The History of the Town and Country of the Town of Galway: From the Earliest Period to the Present Time...* W. Folds, 1820

Harris, A. J.L, and M. Ripepe. *"Regional earthquake as a trigger for enhanced volcanic activity: evidence from MODIS thermal data." Geophysical Research Letters 34.2,* 2007

Harris, R. N. and M. K. McNutt. *"Heat flow on hot spot swells: Evidence for fluid flow." Journal of Geophysical Research: Solid Earth (1978–2012) 112.B3,* 2007

Hartley, R. A., et al. *"Transient convective uplift of an ancient buried landscape." Nature Geoscience 4.8,* 2011, 562-565

Havelock, E. A. *"Preface to Plato. (1963)."* Cambridge, Mass.: Harvard Univer. 1963

Haverty, M. *The History of Ireland, Ancient and Modern: Derived from Our Native Annals, from the Most Recent Researches of Eminent Irish Scholars and Antiquaries, from the State Papers, and from All the Resources of Irish History Now Available...* J. Duffy, 1867

Hill, E. W., M. A. Jobling and D. G. Bradley. *"Y-chromosome variation and Irish origins." Nature 404.6776,* 2000, 351-352

Hitchen, K. *"The geology of the UK Hatton-Rockall margin." Marine and Petroleum Geology 21.8,* 2004, 993-1012

M.J. Hoggard et al. *'Global dynamic topography observations reveal limited influence of large-scale mantle flow.' Nature Geoscience (2016).*

Holford, S. P., et al. *"Regional intraplate exhumation episodes related to plate-boundary deformation." Geological Society of America Bulletin 121.11-12,* 2009, 1611-1628

Holinshed, R. *"The Chronicles of England, Scotland and Ireland."* London, 1587

Hollis, Susan Tower. "Otiose Deities and the Ancient Egyptian Pantheon." *Journal of the American Research Center in Egypt* 35 (1998): 61-72

Hu, A. et al. *"Influence of Bering Strait flow and North Atlantic circulation on glacial sea-level changes." Nature Geoscience 3.2,* 2010, 118-121

Husen, S., et al. *"Changes in geyser eruption behavior and remotely triggered seismicity in Yellowstone National Park produced by the 2002 M 7.9 Denali fault earthquake, Alaska." Geology 32.6,* 2004, 537-540

Huvenne, V. A. I. *"RRS James Cook Cruise 60, 09 May-12 Jun 2011. Benthic habitats and the impact of human activities in Rockall Trough, on Rockall Bank and in Hatton Basin."* 2011

Ireland, R. L., J. F. Poland and F. S. Riley. *Land subsidence in the San Joaquin Valley, California, as of 1980.* U.S. Government Printing Office., 1984

Jacobs, C. L. *"An appraisal of the surface geology and sedimentary processes within SEA7, the UK continental shelf."* 2006

Jagoutz, O. et al. *"Anomalously fast convergence of India and Eurasia caused by double subduction." Nature Geoscience 8.6,* 2015, 475-478

Jakobsson, M. et al. *"Reconstructing the Younger Dryas ice dammed lake in the Baltic Basin: Bathymetry, area and volume." Global and Planetary Change 57.3,* 2007, 355-370

Japsen, P. et al. *"Episodic burial and exhumation in NE Brazil after opening of the South Atlantic." Geological Society of America Bulletin 124.5-6,* 2012, 800-816

Japsen, P. et al. *"Elevated, passive continental margins: Not rift shoulders, but expressions of episodic, post-rift burial and exhumation." Global and Planetary Change 90,* 2012, 73-86.

Jaxybulatov, K., et al. *"A large magmatic sill complex beneath the Toba caldera." Science 346.6209,* 2014, 617-619

Jelinkova, ⊠. AE. "The Shebtiw in the temple at Edfu." *Zeitschrift für Ägyptische Sprache und Altertumskunde* 87.1-2 (1962): 41-54

Jimenez, M-J., and M. Garcı⊠a-Fernández. *"Occurrence of shallow earthquakes following periods of intense rainfall in Tenerife, Canary Islands." Journal of volcanology and geothermal research 103.1,* 2000, 463-468

Jones, S. *The Serpent's Promise: The Bible Retold as Science.* Hachette UK, 2013

Jones, W.H.S. and Ormerod, H.A. English translation of Pausanias' Description of Greece in 4 Volumes. Cambridge, MA, Harvard University Press; London, William Heinemann Ltd. 1918

Josephus, Flavius. - translated by W. Whiston. *"Josephus Complete Works."* Grand Rapids, MI: Kregel, 1960- Reprinted 1976

Kano, A. et al. *"Age constraints on the origin and growth history of a deep-water coral mound in the northeast Atlantic drilled during Integrated Ocean Drilling Program Expedition 307." Geology 35.11,* 2007, 1051-1054

Karlin, R. E. and S. E. B. Abella. *"Paleoearthquakes in the Puget Sound region recorded in sediments from Lake Washington, USA." Science 258.5088,* 1992, 1617-1620

Karmin, M. et al. *"A recent bottleneck of Y chromosome diversity coincides with a global change in culture." Genome Research,* 2015

Keating, G. - translated into English by E. Comyn and P. S. Dinneen *The History of Ireland by Geoffrey Keating* Ex-classics Project, 2009

Keating, G. - translated by Dermod O'Connor *"The General History of Ireland" 1723* Dublin, 1841

Keigwin, L. D. et al. *"Rapid sea-level rise and Holocene climate in the Chukchi Sea." Geology 34.10,* 2006, 861-864

Keigwin, L. D., S. Klotsko, N. Zhao, B. Reilly, L. Giosan, N. W. Driscoll. **Deglacial floods in the Beaufort Sea preceded Younger Dryas cooling.** *Nature Geoscience,* 2018

Kim, Jin-Woo, and Zhong Lu. "Association between localized geohazards in West Texas and human activities, recognized by Sentinel-1A/B satellite radar imagery." *Scientific reports* 8.1 (2018): 4727

Komatsu, G. et al. *"Quaternary paleolake formation and cataclysmic flooding along the upper Yenisei River." Geomorphology 104.3,* 2009, 143-164

Kong, P. et al. *"Moraine dam related to late Quaternary glaciation in the Yulong Mountains, southwest China, and impacts on the Jinsha River." Quaternary Science Reviews 28.27,* 2009, 3224-3235

Kotilainen, A. and K-L. Hutri. *"Submarine Holocene sedimentary disturbances in the Olkiluoto area of the Gulf of Bothnia, Baltic Sea: a case of postglacial palaeoseismicity." Quaternary Science Reviews 23.9,* 2004, 1125-1135

Lacan, M. et al. *"Ancient DNA reveals male diffusion through the Neolithic Mediterranean route." Proceedings of the National Academy of Sciences 108.24,* 2011, 9788-9791

Lagerbäck, R. and M. Sundh *"Early Holocene faulting and paleoseismicity in northern Sweden." National Geological survey of Sweden. Research paper, C 836,* 2008

Lamb, W.R.M. *Plato. Plato in Twelve Volumes, Vol. 9 translated by W.R.M. Lamb.* Cambridge, MA, Harvard University Press; London, William Heinemann Ltd., 1925

Lambeck, K. *"Late Pleistocene and Holocene sea-level change in Greece and south-western Turkey: a separation of eustatic, isostatic and tectonic contributions." Geophysical Journal International 122.3,* 1995, 1022-1044

Laughton, A. S., W. A. Berggren et al., *Initial Reports of the Deep Sea Drilling Project, 12,* U.S. Government Printing Office, 1972

Leschber, C. *"Latin tree names and the European substratum." Studia Linguistica Universitatis Iagellonicae Cracoviensis 129,* 2012, 117-125

Lesko, B. S. *The great goddesses of Egypt.* University of Oklahoma Press, 1999

Leverington, D. W., J. D. Mann, and J. T. Teller. *"Changes in the bathymetry and volume of glacial Lake Agassiz between 11,000 and 9300 14C yr BP." Quaternary Research 54.2,* 2000, 174-181

Leverington, D. W., J. D. Mann and J. T. Teller. *"Changes in the bathymetry and volume of glacial Lake Agassiz between 9200 and 7700 14C yr BP." Quaternary Research 57.2,* 2002, 244-252

Li, C. et al. *"Abrupt climate shifts in Greenland due to displacements of the sea ice edge."* *Geophysical Research Letters 32.19*, 2005

Li, C. and D. S. Battisti. *"Reduced Atlantic storminess during Last Glacial Maximum: Evidence from a coupled climate model." Journal of Climate 21.14*, 2008, 3561-3579

Li, Ji, et al. *"The latest straight-tusked elephants (Palaeoloxodon)?"Wild elephants" lived 3000 years ago in North China." Quaternary International 281*, 2012, 84-88

Lioubimtseva, E. U., S. P. Gorshkov, and J. M. Adams. *"A giant Siberian lake during the last glacial: evidence and implications." Unpublished paper. http://www. esd. ornl.gov/ projects/qen/lake. Html (2001)*, Accessed on 08 October 2014

Lister, A. M. *"Late-glacial mammoth skeletons (Mammuthus primigenius) from Condover (Shropshire, UK): anatomy, pathology, taphonomy and chronological significance." Geological Journal 44.4*, 2009, 447-479

Lohne, Ø. S. et al. *"Calendar year age estimates of Allerød–Younger Dryas sea-level oscillations at Os, western Norway." Journal of Quaternary Science 19.5*, 2004, 443-464

Lôugas, L., P. Ukkonen, and H. Jungner. *"Dating the extinction of European mammoths: new evidence from Estonia." Quaternary Science Reviews 21.12*, 2002, 1347-1354

Maca-Meyer, N. et al. *"Ancient mtDNA analysis and the origin of the Guanches." European journal of human genetics 12.2*, 2004, 155-162

Macaulay, G.C. (Translator) *The History of Herodotus, by Herodotus.*

Maccaferri, F. et al. *"Off-rift volcanism in rift zones determined by crustal unloading." Nature Geoscience*, 2014

Manga, M. and E. Brodsky. *"Seismic triggering of eruptions in the far field: volcanoes and geysers." Annu. Rev. Earth Planet. Sci 34*, 2006, 263-291

Mann, J. D. et al. *"The volume and paleobathymetry of glacial Lake Agassiz." Journal of Paleolimnology 22.1*, 1999, 71-80

Margold, M. et al. *"Glacial Lake Vitim, a 3000-km3 outburst flood from Siberia to the Arctic Ocean." Quaternary Research 76.3*, 2011, 393-396

Markova, A. K., et al. "New data on changes in the European distribution of the mammoth and the woolly rhinoceros during the second half of the Late Pleistocene and the early Holocene." *Quaternary International* 292 (2013): 4-14.

Martin, P. S. and A.J. Stuart. *"Mammoth extinction: two continents and Wrangel Island." Radiocarbon, Vol. 37.1, 1995*, 7-10

Marzocchi, W.. *"Remote seismic influence on large explosive eruptions." Journal of Geophysical Research: Solid Earth (1978–2012) 107.B1 EPM-6.*, 2002, EPM 6-1–EPM 6-7

Masseti, M. *"On the Pleistocene occurrence of Elephas (Palaeoloxodon) antiquus in the Tuscan Archipelago, northern Tyrrhenian Sea (Italy)." Hystrix, The Italian Journal of Mammalogy 5.1-2*, 1994

Masse, W. B. et al. *"Exploring the nature of myth and its role in science." Geological Society, London, Special Publications 273.1*, 2007, 9-28

McClain, Brett, 2011, Cosmogony (Late to Ptolemaic and Roman Periods). In Jacco Dieleman, Willeke Wendrich (eds.), UCLA Encyclopedia of Egyptology, Los Angeles.

McLauchlan, Thomas. "The Dean of Lismore's Book." (1862).

McGee, T. D'Arcy. *A Popular History of Ireland: From the Earliest Period to the Emancipation of the Catholics (Complete). Vol. 1.*, Library of Alexandria, 1869

McGrail, S. *Ancient Boats in North-West Europe: The archaeology of water transport to AD 1500*. Routledge, 2014,

McMichael, A. J. *"Insights from past millennia into climatic impacts on human health and survival."* Proceedings of the National Academy of Sciences 109.13, 2012, 4730-4737

McMillan, M. et al. *"Three-dimensional mapping by CryoSat-2 of subglacial lake volume changes."* Geophysical Research Letters 40.16, 2013, 4321-4327

McNutt, M. and A. Bonneville. *"A shallow, chemical origin for the Marquesas Swell."* Geochemistry, Geophysics, Geosystems 1.6, 2000

Meinsen, J. et al. *"Middle Pleistocene (Saalian) lake outburst floods in the Münsterland Embayment (NW Germany): impacts and magnitudes."* Quaternary Science Reviews 30.19, 2011, 2597-2625

Meyer, H. et al. *"Permafrost evidence for severe winter cooling during the Younger Dryas in northern Alaska."* Geophysical Research Letters 37.3, 2010

Meyer, R., J. van Wijk, and L. Gernigon. *"The North Atlantic Igneous Province: A review of models for its formation."* Geological Society of America Special Papers 430, 2007, 525-552.

Mienis, F., et al. *"Sediment accumulation on a cold-water carbonate mound at the Southwest Rockall Trough margin."* Marine Geology 265.1, 2009, 40-450

Miller, M. S. and T. W. Becker. *"Reactivated lithospheric-scale discontinuities localize dynamic uplift of the Moroccan Atlas Mountains."* Geology 42.1, 2014, 35-38

Missenard, Y. et al. *"Crustal versus asthenospheric origin of relief of the Atlas Mountains of Morocco."* Journal of Geophysical Research: Solid Earth (1978–2012) 111.B3, 2006

Mitrovica, J. X., and G. A. Milne. *"On the origin of late Holocene sea-level highstands within equatorial ocean basins."* Quaternary Science Reviews 21.20, 2002, 2179-2190

Monaghan, P. *The encyclopedia of Celtic mythology and folklore*. Infobase Publishing, 2009

Moore, T. *The history of Ireland*. Baudry, 1837

Mörner, N-A. *"The Fenris Wolf in the Nordic Asa creed in the light of palaeoseismics."* Geological Society, London, Special Publications 273.1, 2007, 117-119

Mörner, N-A. *"Liquefaction and varve deformation as evidence of paleoseismic events and tsunamis. The autumn 10,430 BP case in Sweden."* Quaternary Science Reviews 15.8, 1996, 939-948

Morton, A. C. et al. *"Detrital zircon age constraints on the provenance of sandstones on Hatton Bank and Edoras Bank, NE Atlantic."* Journal of the Geological Society 166.1, 2009, 137-146.

Murray, A.T. (Translator) *Homer. The Odyssey with an English translation in two volumes*. Cambridge, MA., Harvard University Press; London, William Heinemann Ltd., 1919

Murton, J. B. et al. *"Identification of Younger Dryas outburst flood path from Lake Agassiz to the Arctic Ocean."* Nature 464.7289, 2010, 740-743

Neish, J. K. *"Seismic structure of the Hatton–Rockall area: an integrated seismic/modelling study from composite datasets."* Geological Society, London, Petroleum Geology Conference series. Vol. 4. Geological Society of London, 1993

Nunn, P. D. *"On the convergence of myth and reality: examples from the Pacific Islands."* The Geographical Journal 167.2, 2001, 125-138

Nunn, P. D. et al. *"Vanished islands in Vanuatu: new research and a preliminary geohazard assessment."* Journal of the Royal Society of New Zealand 36.1, 2006, 37-50

Nunn, P, *The Edge of Memory: The Geology of Folk Tales and Climate Change*, Bloomsbury Publishing Plc, 2018

Nunn, P. D. and N. J. Reid *"Aboriginal Memories of Inundation of the Australian Coast Dating from More than 7000 Years Ago" Australian Geographer, DOI: 10.1080/00049182.2015.1077539,* 2015

O'Connell, M., C. C. Huang, and U. Eicher. *"Multidisciplinary investigations, including stable-isotope studies, of thick Late-glacial sediments from Tory Hill, Co. Limerick, western Ireland." Palaeogeography, Palaeoclimatology, Palaeoecology 147.3,* 1999, 169-208

O'Flaherty, R. - Translated by the Revd. J. Hely, *"Ogygia, or, a chronological account of Irish events:: collected from very ancient documents "* 1793

Okrostsvaridze, A. and D. Bluashvili. *"Mythical "Gold Sands" of Svaneti (Greater Caucasus, Georgia): Geological Reality and Gold Mining Artefacts." Bull. Georg. Natl. Acad. Sci 4.2,* 2010,

Okrostsvaridze, A., N. Gagnidze, and K. Akimidze. *"A modern field investigation of the mythical "gold sands" of the ancient Colchis Kingdom and "Golden Fleece" phenomena." Quaternary International,* 2014

Olesen, O. et al. *"Neotectonic deformation in Norway and its implications: a review." Norsk Geologisk Tidsskrift 84.1,* 2004, 3 – 34

Orlova, L. A., et al. *"Lugovskoe, western Siberia: A possible extra-Arctic mammoth refugium at the end of the Late Glacial." Radiocarbon 46.1,* 2004, 363-368

Oxford English Dictionary, The Shorter - on historical principles, (Vol. 1 A-M). 2002

Pascal, C. and O. Olesen. *"Are the Norwegian mountains compensated by a mantle thermal anomaly at depth?" Tectonophysics 475.1,* 2009, 160-168

Paschou, P. et al. *"Maritime route of colonization of Europe." Proceedings of the National Academy of Sciences 111.25,* 2014, 9211-9216

Pearson, I. and D. G. Jenkins. *"Unconformities in the Cenozoic of the North-East Atlantic." Geological Society, London, Special Publications 21.1,* 1986, 79-86

Peters, J. and K. Schmidt. *"Animals in the symbolic world of Pre-Pottery Neolithic Göbekli Tepe, south-eastern Turkey: a preliminary assessment." Anthropozoologica 39.1,* 2004, 179-218.

Petford, N. et al. *"Granite magma formation, transport and emplacement in the Earth's crust." Nature 408.6813,* 2000, 669-673

Pfeffer, Julia, et al. "Decoding the origins of vertical land motions observed today at coasts." *Geophysical Journal International* 210.1 (2017): 148-165

Pirlet, H. et al. *"Diagenetic formation of gypsum and dolomite in a cold-water coral mound in the Porcupine Seabight, off Ireland." Sedimentology 57.3,* 2010, 786-805

Plutarch *Moralia, Vol XII, 26 Concerning the Face Which Appears in the Orb of the Moon*

Poland, J. F. *"Land subsidence in the San Joaquin Valley, California, and its effect on estimates of ground-water resources." Internat. Assoc. of Sci. Hydrology, Publ 52,* 1960, 324-335

Poore, H. R., N. White, and S. Jones. *"A Neogene chronology of Iceland plume activity from V-shaped ridges." Earth and Planetary Science Letters 283.1,* 2009, 1-13

Poore, H., N. White, and J. Maclennan. *"Ocean circulation and mantle melting controlled by radial flow of hot pulses in the Iceland plume." Nature Geoscience 4.8,* 2011, 558-561

Pritchard, M. E., et al. *"Subsidence at southern Andes volcanoes induced by the 2010 Maule, Chile earthquake." Nature Geoscience 6.8,* 2013, 632-636

Pseudo Scymnus', or Pausanias of Damascus', *Circuit of the Earth* (unedited translation by Brady Kiesling from the Muller text *(Geographi Graeci Minores)* of 1857) accessed at https://topostext.org/work/130

Rahmstorf, S. *"Timing of abrupt climate change: A precise clock." Geophysical Research Letters 30.10,* 2003,

Rappenglück, B. et al. *"The fall of Phaethon: a Greco-Roman geomyth preserves the memory of a meteorite impact in Bavaria (south-east Germany)." Antiquity 84.324,* 2010, 428-439

Rasmussen, S. O. et al. *"A new Greenland ice core chronology for the last glacial termination." Journal of Geophysical Research: Atmospheres (1984–2012) 111.D6,* 2006

Rasmussen, T. L., E. Thomsen, and M. Moros. *"North Atlantic warming during Dansgaard-Oeschger events synchronous with Antarctic warming and out-of-phase with Greenland climate." Scientific Reports* **6**, *20535,* 2016

Raukas, A. and W. Stankowski. *"On the age of the Kaali craters, Island of Saaremaa, Estonia." Baltica 24.1,* 2011, 37-44

Rawlinson, G. (Translator) *The History of Herodotus by Herodotus, Book 1,* Clio

Reuther, A U., et al. *"Constraining the timing of the most recent cataclysmic flood event from ice-dammed lakes in the Russian Altai Mountains, Siberia, using cosmogenic in situ 10Be." Geology 34.11,* 2006, 913-916

Reymond, Eve AE. *The mythical origin of the Egyptian temple. Manchester University Press,* 1969

Rickers, F., A. Fichtner and J. Trampert. *The Iceland–Jan Mayen plume system and its impact on mantle dynamics in the North Atlantic region: Evidence from full-waveform inversion." Earth and Planetary Science Letters 367,* 2013, 39-51

Roberts, D. G. *APPENDIX II SITE SURVEY IN Hatton-Rockall BASIN (Sites 116 and 117)*

Roberts, D. G. *"Marine geology of the Rockall Plateau and Trough." Philosophical Transactions of the Royal Society of London. Series A, Mathematical and Physical Sciences 278.1285,* 1975, 447-509

Rodolfo, K. S. and J. V. Umbal. *"A prehistoric lahar-dammed lake and eruption of Mount Pinatubo described in a Philippine aborigine legend." Journal of Volcanology and Geothermal Research 176.3,* 2008, 432-437

Rolfe, J. C. (translator) *Ammianus Marcellinus. Loeb Classical Library. 3 vols.* Cambridge, Mass., 1935-9

Rosser, Z. H. et al. *"Y-chromosomal diversity in Europe is clinal and influenced primarily by geography, rather than by language." The American Journal of Human Genetics 67.6,* 2000, 1526-1543

Rovelli, C. *http://www.edge.org/conversation/a-philosophy-of-physics,* 30 May 2012

Rudge, J. F., et al. *"A plume model of transient diachronous uplift at the Earth's surface." Earth and Planetary Science Letters 267.1,* 2008, 146-160

Sayago-Gil, M, D. Long, K. Hitchen, V. Díaz-del-Río, L.M. Fernández-Salas and P. Durán-Muñoz *Geomorphology of the western slope of Hatton Bank (Rockall Plateau, NE Atlantic Ocean) revealed by multibeam bathymetry and high-resolution seismic data:*

control by bottom current regime nora.nerc.ac.uk/9492/1/Sayago-Gil_et_al_v1.pdf - NERC Open Access Research Archive (NORA), 2009 - Accessed on 20 October 2014

Schmidt, K. *"Göbekli Tepe–the Stone Age Sanctuaries. New results of ongoing excavations with a special focus on sculptures and high reliefs." Documenta Praehistorica 37*, 2010, 239-256.

Schrijver, P. *"Varia V. Non-Indo-European Surviving in Ireland in the First Millennium AD." Ériu*, 2000, 195-199.1

Service, A. *Lost worlds.* Arco Pub., 1981

Shimizu, H. *"Past, Present and Future of the Pinatubo Aetas: At the Crossroads of Socio-Cultural Disintegration." After the Eruption: Pinatubo Aetas at the Crisis of their Survival*, 1992, 1-49

Shuckburgh, Evelyn S., ed. *The Histories of Polybius: Translated from the Text of F. Hultsch.* Vol. 2. from http://www.gutenberg.org/ebooks/44126 (Accessed on 7 September 2018)

Smith, Dr. W. *"Dictionary of Ancient Geography."*

Smith, L. K., R. S. White, N.J. Kusznir and iSIMM Team *"Structure of the Hatton Basin and adjacent continental margin."*

in: Dore, A. G. & Vining, B. A. (eds) *Petroleum Geology: North-West Europe and Global Perspectives—Proceedings of the 6th Petroleum Geology Conference*, 2005, 947–956

Smyth, J. and R. P. Evershed. *"Milking the megafauna: Using organic residue analysis to understand early farming practice." Environmental Archaeology*, 2015

Sternai, P., L. Caricchi, S. Castelltort, and J.-D. Champagnac , *Deglaciation and glacial erosion: a joint control on magma productivity by continental unloading, (Abstract) Geophys. Res. Lett., 43*, 2016

Stewart, J. R., et al. *"Refugia revisited: individualistic responses of species in space and time." Proceedings of the Royal Society B: Biological Sciences 277.1682*, 2010, 661-671

Stoker, M. S., et al. *"Neogene evolution of the Atlantic continental margin of NW Europe (Lofoten Islands to SW Ireland): anything but passive." Petroleum Geology: North West Europe and Global Perspectives: Proceedings of the 6th Conference.. Geological Society of London*, 2005

Stoker, M. S., et al. *"Eocene post-rift tectonostratigraphy of the Rockall Plateau, Atlantic margin of NW Britain: linking early spreading tectonics and passive margin response." Marine and Petroleum Geology 30.1*, 2012, 98-125

Strabo. *The Geography of Strabo, Vol. I of the Loeb Classical Library* Loeb Classical Library, 1917

Strabo. - translated by H. L. Jones *Geographica or The Geography of Strabo (Loeb Classical Library edition Vol. 2 of 8)* Harvard University Press, 1923

Stratospheric, A. and D. Bluashvili. *"Mythical "Gold Sands" of Svaneti (Greater Caucasus, Georgia): Geological Reality and Gold Mining Artefacts." Bull. Georg. Natl. Acad. Sci 4.2*, 2010

Stuart, A. J. *"The vertebrates of the last cold stage in Britain and Ireland." Philosophical Transactions of the Royal Society of London. B, Biological Sciences 280.972*, 1977, 295-312

Stuart, A. J., et al. *"The latest woolly mammoths (Mammuthus primigenius Blumenbach) in Europe and Asia: a review of the current evidence." Quaternary Science Reviews 21.14*, 2002, 1559-1569

Stuart, A. J. *"The extinction of woolly mammoth (Mammuthus primigenius) and straight-tusked elephant (Palaeoloxodon antiquus) in Europe."* Quaternary International 126, 2005, 171-177

Sutinen, R. et al. *"Airborne LiDAR detection of postglacial faults and Pulju moraine in Palojärvi, Finnish Lapland."* Global and Planetary Change 115, 2014, 24-32

Sutinen, R., M. Piekkari, and M. Middleton. *"Glacial geomorphology in Utsjoki, Finnish Lapland proposes Younger Dryas fault-instability."* Global and Planetary Change 69.1, 2009, 16-28

Takada, Y. and Y. Fukushima. *"Volcanic subsidence triggered by the 2011 Tohoku earthquake in Japan."* Nature Geoscience 6.8, 2013, 637-641

Talwani, P. *"On the nature of reservoir-induced seismicity."* Pure and applied geophysics 150.3-4, 1997, 473-492

Tambets, K. et al. *"The western and eastern roots of the Saami⎯the story of genetic "outliers" told by mitochondrial DNA and Y chromosomes."* The American Journal of Human Genetics 74.4, 2004, 661-682.

Tang, Liujuan, et al. *"Direct energy estimation of the 2011 Japan tsunami using deep-ocean pressure measurements."* Journal of Geophysical Research Oceans 117.C8 (2012)

Taylor, A. E. *"Plato, the Man and His Work. 7th⎯ed.",* London: Methuen, 1960

Taylor, K. C., et al. *"The Holocene-Younger Dryas transition recorded at Summit, Greenland."* Science 278.5339, 1997, 825-827

Teller, J. T., and D. W. Leverington. *"Outbursts from Lake Agassiz and their possible impact on coastal environments."* Conference abstracts: Environmental Catastrophes and Recoveries in the Holocene. Uxbridge, UK: Brunel University, 2002

Teller, J. T., D. W. Leverington, and J. D. Mann. *"Freshwater outbursts to the oceans from glacial Lake Agassiz and their role in climate change during the last deglaciation."* Quaternary Science Reviews 21.8, 2002, 879-887

Thornalley, D. J. R., et al. *"A warm and poorly ventilated deep Arctic Mediterranean during the last glacial period."* Science 349.6249, 2015, 706-710

Torroni, A. et al. *"A signal, from human mtDNA, of postglacial recolonization in Europe."* The American Journal of Human Genetics 69.4, 2001, 844-852

Ukkonen, P., et al. "Woolly mammoth (Mammuthus primigenius Blum.)and its environment in northern Europe during the last glaciation." *Quaternary Science Reviews* 30.5-6 (2011): 693-712.

Utkucu, M. *"Implications for the water level change triggered moderate (M≥ 4.0) earthquakes in Lake Van basin, Eastern Turkey."* Journal of seismology 10.1, 2006, 105-117

van der Land, C. et al. *"Carbonate mound development in contrasting settings on the Irish margin."* Deep Sea Research Part II: Topical Studies in Oceanography 99, 2014, 297-306

Van der Plicht, J., et al. *"New Holocene refugia of giant deer (Megaloceros giganteus Blum.) in Siberia: updated extinction patterns."* Quaternary Science Reviews 114, 2015, 182-188

van Hinsbergen, D. J. J, et al. *"Acceleration and deceleration of India-Asia convergence since the Cretaceous: Roles of mantle plumes and continental collision."* Journal of Geophysical Research: Solid Earth (1978–2012) 116.B6, 2011

Vartanyan, S. L., et al. *"Radiocarbon dating evidence for mammoths on Wrangel Island, Arctic Ocean, until 2000 BC."* Radiocarbon 37.1, 1995, 1-6

Vennemann, T. *"Atlantis Semitica: structural contact features in Celtic and English."* *Amsterdam Studies in the theory and history of Linguistic Science Series 4*, 1999, 351-370

Veski, S., et al. *"The physical and social effects of the Kaali meteorite impact⊠a review."* *Comet/asteroid impacts and human society*, Springer Berlin Heidelberg, 2007, 265-275

Vinson, D. A. *"The Western Sea: Atlantic History before Columbus."* *Northern Mariner 10.3*, 2000, 1-14

Vitaliano, D. B., *Geomythology. Journal of the Folklore Institute, Vol. 5, No. 1*, 1968

Vitaliano, D. B. *"Geomythology: geological origins of myths and legends."* *Geological Society, London, Special Publications 273.1*, 2007, 1-7

Walter, T. R. *"How a tectonic earthquake may wake up volcanoes: Stress transfer during the 1996 earthquake–eruption sequence at the Karymsky Volcanic Group, Kamchatka."* *Earth and Planetary Science Letters 264.3*, 2007, 347-359

Watt, S. F.L, D. M. Pyle, and T. A. Mather. *"The influence of great earthquakes on volcanic eruption rate along the Chilean subduction zone."* *Earth and Planetary Science Letters 277.3*, 2009, 399-407

Watts, A. B., B. C. Schreiber, and D. Habib. *"Dredged rocks from Hatton Bank, Rockall Plateau."* *Journal of the Geological Society 131.6*, 1975, 639-646

Wauthier, C., B. Smets, and D. Keir , *"Diking-induced moderate- magnitude earthquakes on a youthful rift border fault: The 2002 Nyiragongo- Kalehe sequence, D.R. Congo"* *Geochem. Geophys. Geosyst., 16*, 2015, 4280– 4291

Weidle, C. and V. Maupin. *"An upper-mantle S-wave velocity model for Northern Europe from Love and Rayleigh group velocities."* *Geophysical Journal International 175.3*, 2008, 1154-1168

Wentz, W. Y. *"Evans. The Fairy-Faith in Celtic Countries."* 1911

White, R. S. and L. K. Smith. *"Crustal structure of the Hatton and the conjugate east Greenland rifted volcanic continental margins, NE Atlantic."* *Journal of Geophysical Research: Solid Earth (1978–2012) 114.B2*, 2009

White, R. and D. McKenzie. *"Magmatism at rift zones: the generation of volcanic continental margins and flood basalts."* *Journal of Geophysical Research: Solid Earth (1978–2012) 94.B6*, 1989, 7685-7729

Wickham-Jones, C. R., and S. Dawson. *"The scope of Strategic Environmental Assessment of North Sea Area SEA7 with regard to prehistoric and early historic archaeological remains."* *www.offshore-sea.org.uk/site/index. Php*, 2006

Wienberg, C. et al. *"Glacial cold-water coral growth in the Gulf of Cádiz: Implications of increased palaeo-productivity."* *Earth and Planetary Science Letters 298.3*, 2010, 405-416

Xiaoming Liu, Y. Fu. *"Exploring population size changes using SNP frequency spectra."* *Nature Genetics 47*, 2015, 555-569

Zamelczyk, K. et al. *"Paleoceanographic changes and calcium carbonate dissolution in the central Fram Strait during the last 20ka."* *Quaternary Research 78.3*, 2012, 405-416

Zdanowicz, C. M., G. A. Zielinski, and M. S. Germani. *"Mount Mazama eruption: Calendrical age verified and atmospheric impact assessed."* *Geology 27.7*, 1999, 621-624

Zhang, Kui, et al. "Monitoring ground surface deformation over the North China Plain using coherent ALOS PALSAR differential interferograms." *Journal of Geodesy* 87.3 (2013): 253-265.

Zhirov, N. *Atlantis: Atlantology: Basic Problems.* University Press of the Pacific, 2001

Zielinski, G. A. *"Use of paleo-records in determining variability within the volcanism–climate system."* Quaternary Science Reviews 19.1, 2000, 417-438

Zielinski, G. A. et al. *"A 110,000-yr record of explosive volcanism from the GISP2 (Greenland) ice core."* Quaternary Research 45.2, 1996, 109-118

Summary and Conclusions, p. 1188 of the Deep Sea Drilling Project, http://www.deepseadrilling.org/12/dsdp_toc.htm, 2007

Southern Methodist University. "Radar images show large swath of Texas oil patch is heaving and sinking at alarming rates." ScienceDaily. ScienceDaily,21 March 2018. www.sciencedaily.com/releases/2018/03/180321110855.htm

Dartmouth College. "Prehistoric climate shift linked to cosmic impact." ScienceDaily. ScienceDaily, 2 September 2013, www.sciencedaily.com/releases/2013/09/130902162707.htm

Universität Bonn. "Devastating long-distance impact of earthquakes." ScienceDaily. ScienceDaily, 23 July 2013. www.sciencedaily.com/releases/2013/07/130723073957.htm

Massachusetts Institute of Technology. "Mystery of India's rapid move toward Eurasia 80 million years ago explained." ScienceDaily. ScienceDaily, 4 May 2015. www.sciencedaily.com/releases/2015/05/150504120815.htm

Helmholtz Centre Potsdam - GFZ German Research Centre for Geosciences. "Magma pancakes beneath Indonesia's Lake Toba: Subsurface sources of mega-eruptions." ScienceDaily. ScienceDaily, 30 October 2014. www.sciencedaily.com/releases/2014/10/141030142024.htm

Rice University. "Connection between Hawaii's dueling volcanoes explained." ScienceDaily. ScienceDaily, 23 October 2012 www.sciencedaily.com/releases/2012/10/121023134810.htm

Helmholtz Centre Potsdam - GFZ German Research Centre for Geosciences. "Off-rift volcanoes explained: Crustal unloading." ScienceDaily. ScienceDaily, 23 March 2014. www.sciencedaily.com/releases/2014/03/140323151958.htm

Wiley-Blackwell. "Mammoths Survived In Britain Until 14,000 Years Ago, New Discovery Suggests." ScienceDaily. ScienceDaily, 18 June 2009. www.sciencedaily.com/releases/2009/06/090617201758.htm

University of Texas Health Science Center at Houston. "New population genetics model could explain Finn, European genetic differences." ScienceDaily. ScienceDaily, 11 May 2015. www.sciencedaily.com/releases/2015/05/150511095358.htm

University of Cambridge. "Increase in volcanic eruptions at the end of the ice age caused by melting ice caps and erosion." ScienceDaily. ScienceDaily, 1 February 2016. <www.sciencedaily.com/releases/2016/02/160201141731.htm>

CAGE - Center for Arctic Gas Hydrate, Climate and Environment. "'Ice age blob' of warm ocean water discovered south of Greenland." ScienceDaily. ScienceDaily, 19 February 2016. <www.sciencedaily.com/releases/2016/02/160219134816.htm>

University of Cambridge. "Map of flow within the Earth's mantle finds the surface moving up and down 'like a yo-yo'." ScienceDaily. ScienceDaily, 9 May 2016. <www.sciencedaily.com/releases/2016/05/160509115116.htm>

Web Resources Referenced

URL	Date last accessed
http://www.edge.org/conversation/a-philosophy-of-physics	22 October 2014
https://en.wikipedia.org/wiki/Gobekli_Tepe	29 March 2016
https://en.wikipedia.org/wiki/Natufian_culture	29 March 2016
http://en.wikipedia.org/wiki/Quran	26 October 2014
http://en.wikipedia.org/wiki/Hafiz_%28Quran%29	26 October 2014
https://en.wikipedia.org/wiki/Geomythology	
https://en.wikipedia.org/wiki/Euhemerus	
http://en.wikipedia.org/wiki/Mount_Mazama	24 May 2015
http://en.wikipedia.org/wiki/Lake_Euramo	26 October 2014
http://www.thefreelibrary.com/Origin+legends+and+linguistic+relationships.-a019175999	26 October 2014
http://en.wikipedia.org/wiki/Prose_Edda	26 October 2014
http://en.wikipedia.org/wiki/Solon	25 October 2014
http://en.wikipedia.org/wiki/Sais,_Egypt	25 October 2014
http://en.wikipedia.org/wiki/Neith	25 October 2014
http://en.wikipedia.org/wiki/Sais,_Egypt	25 October 2014
http://en.wikipedia.org/wiki/Sonchis_of_Sais	25 October 2014
http://www.iep.utm.edu/plato/	26 October 2014
http://www.ancient.eu.com/plato/	26 October 2014
http://en.wikipedia.org/wiki/Plato	26 October 2014
http://en.wikipedia.org/wiki/Homer	
http://penelope.uchicago.edu/Thayer/E/Roman/Texts/Strabo/1B1*.html	21 September 2014
https://en.wikipedia.org/wiki/Rockall	11 August 2015
https://en.wikipedia.org/wiki/Olaus_Rudbeck	13 August 2015
https://en.wikipedia.org/wiki/Mecklenburg	13 August 2015
http://www.mareano.no/en/maps/mareano_en.html	13 August 2015
http://en.wikipedia.org/wiki/Nymph	19 September 2014
http://www.theoi.com/Titan/Hesperides.html	19 September 2014

URL	Date accessed
http://en.wikipedia.org/wiki/Hesperides	19 September 2014
http://www.perseus.tufts.edu/Herakles/apples.html	19 September 2014
http://penelope.uchicago.edu/Thayer/E/Roman/Texts/Diodorus_Siculus/4B*.html	19 September 2014
https://translate.google.co.za/?hl=en&tab=wT#el/en/%-CE%BC%CE%B7%CE%BB%CE%BF%CF%82	02 February 2015
https://translate.google.co.za/?hl=en&tab=wT#el/en/%-CE%95%E1%BD%94%CE%BC%CE%B7%-CE%BB%CE%BF%CF%82.	02 February 2015
https://en.wikipedia.org/wiki/Hecataeus_of_Abdera	31 August 2016
https://en.wikipedia.org/wiki/Brasil_(mythical_island)	09 April 2018
http://en.wikipedia.org/wiki/Avalon	09 June 2014
http://en.wikipedia.org/wiki/Emain_Ablach	09 June 2014
http://www.sacred-texts.com/neu/eng/vm/vmeng.htm	09 June 2014
http://www.maryjones.us/jce/otherworld.html	09 June 2014
http://www.gutenberg.org/browse/authors/p#a93	29 March 2016
http://www.perseus.tufts.edu	29 March 2016
http://goo.gl/AUatI1 (Original un-shortened: http://www.perseus.tufts.edu/hopper/text?doc=Perseus%-3Atext%3A1999.01.0164%3Aletter%3D7%-3Asection%3D344d)	6 February 2015
http://goo.gl/IEyITN (Original un-shortened: http://www.perseus.tufts.edu/hopper/text?doc=Perseus%-3Atext%3A1999.01.0174%3Atext%-3DPhaedrus%3Asection%3D276d)	6 February 2015
https://en.wikipedia.org/wiki/Republic_%28Plato%29	31 January 2016
http://en.wikipedia.org/wiki/Stadion_%28unit_of_length%29	29 March 2016
https://en.wikipedia.org/wiki/Hecataeus_of_Miletus	17 April 2018
https://en.wikipedia.org/wiki/Colaeus	
https://en.wikipedia.org/wiki/Tartessos	
https://en.wikipedia.org/wiki/Periplus_of_Pseudo-Scylax	
http://penelope.uchicago.edu/aelian/varhist5.xhtml	10 July 2018
https://en.wikipedia.org/wiki/Pseudo-Aristotle	
https://en.wikipedia.org/wiki/On_the_Universe	
https://en.wikipedia.org/wiki/	
https://en.wikipedia.org/wiki/Oceanus	
https://en.wikipedia.org/wiki/Stesichorus	

https://en.wikipedia.org/wiki/Eratosthenes	
https://en.wikipedia.org/wiki/File:Mappa_di_Eratostene.jpg	
https://en.wikipedia.org/wiki/Pytheas	
http://platoproject.gr/project/the-importance-of-accurate-translation/	
https://el.wikipedia.org/wiki/%CE%A0%CE%AD%-CE%BB%CE%B1%CE% B3%CE%BF%CF%82	
http://biblehub.com/greek/2281.htm	
http://etymonline.com/index.php?term=pan-.	
https://en.wikipedia.org/wiki/Pan	
http://classics.mit.edu/Aristotle/meteorology.2.ii.html	18 April 2018
https://en.wikipedia.org/wiki/Pytheas#cite_note-10	
https://en.wikipedia.org/wiki/Cassiterides	
https://en.wikipedia.org/wiki/Carthage	
https://en.wikipedia.org/wiki/Sicilian_Wars	
https://en.wikipedia.org/wiki/Spartel	
http://atlantipedia.ie/samples/continent/	09 April 2018
https://en.wikipedia.org/wiki/Continent#The_word_continent	17 April 2018
https://en.wiktionary.org/wiki/continent	17 April 2018
http://www.yourguidetoitaly.com/origin-of-the-name-italy.html	
http://atlantipedia.ie/samples/italy-n/	05 March 2018
http://atlantipedia.ie/samples/size-of-atlantis/	09 April 2018
http://atlantipedia.ie/samples/asia/	17 April 2018
http://atlantipedia.ie/samples/libya/	17 April 2018
https://en.wikipedia.org/wiki/Rioni_River	05 March 2018
https://en.wikipedia.org/wiki/Kerch_Strait	05 March 2018
https://en.wikipedia.org/wiki/Diodorus_Siculus	05 March 2018
http://ports.com	
perseus.tufts.edu - http://goo.gl/vUX3my (Original un-shortened: http://www.perseus.tufts.edu/hopper/morph?l=pro\&la=greek&can=pro\0&prior=ga\r&d=Perseus:text:1999.01.0179:text=Tim.:section=24e&i=1)	13 March 2015
http://www.etymonline.com/index.php?term=pro-	13 March 2015

http://www.etymonline.com/index.php?term=per&allowed_in_frame=0	13 March 2015
http://logeion.uchicago.edu/index.html#%CF%80%-CF%81%CF%8C	13 March 2015
http://en.wikipedia.org/wiki/Temperate_climate	30 October 2014
https://en.wikipedia.org/wiki/Norse_colonization_of_North_America	10 September 2018
http://maps.ngdc.noaa.gov/viewers/bathymetry/	24 October 2014
http://en.wikipedia.org/wiki/File:World_geologic_provinces.jpg	26 September 2014
http://en.wikipedia.org/wiki/Subaerial	26 September 2014
http://en.wikipedia.org/wiki/Nautical_miles	19 October 2014
http://en.wikipedia.org/wiki/Republic_of_Ireland	19 October 2014
http://en.wikipedia.org/wiki/Iceland	19 October 2014
http://en.wikipedia.org/wiki/Geography_of_the_United_Kingdom	19 October 2014
http://en.wikipedia.org/wiki/List_of_mountains_and_hills_of_the_United_Kingdom	24 October 2014
http://en.wikipedia.org/wiki/Carrauntoohil	24 October 2014
http://www.ngdc.noaa.gov/mgg/topo/globega2.html	24 October 2014
http://www.ngdc.noaa.gov/mgg/topo/pictures/GLOBALsealevelsm.jpg	24 October 2014
http://www.seismicatlas.org/entity?id=bc82829e-e629-4036-a7d8-aa7015ded422	24 October 2014
http://www.deepseadrilling.org/12/dsdp_toc.htm	19 September 2014
http://thinkingdeepblue.blogspot.com/	17 October 2014
http://www.prairie.illinois.edu/shilts/gallery/shilts-tyrrell.shtml	17 October 2014
http://www.eu-hermione.net/rockall-hatton-basin	17 October 2014
https://en.wikipedia.org/wiki/Lake_Agassiz	4 September 2018
http://en.wikipedia.org/wiki/List_of_prehistoric_lakes	29 March 2016
https://en.wikipedia.org/wiki/List_of_reservoirs_by_volume	16 August 2018
https://en.wikipedia.org/wiki/Lake_Kariba	4 September 2018
https://www.herald.co.zw/quake-rattles-kariba/	31 August 2018
https://earthobservatory.nasa.gov/images/87485/the-decline-of-lake-kariba	31 August 2018

http://see-atlas.leeds.ac.uk:8080/entity?id=bc82829e-e629-4036-a7d8-aa7015ded422	29 March 2016
http://www.seismicatlas.org/entity?id=6a73b36f-32cd-4014-ac02-20ebf16f63bf	20 October 2014
http://www.bbc.com/news/uk-scotland-29539119	12 August 2015
https://en.wikipedia.org/wiki/Climate	23 June 2015
http://www.ipcc.ch/publications_and_data/ar4/wg1/en/faq-1-1.html	23 June 2015
https://en.wikipedia.org/wiki/LOWERN	23 June 2015
http://www.climateandweather.net/global-warming/factors-that-influence-climate.html	23 June 2015
http://www.ncdc.noaa.gov/paleo/milankovitch.html	23 June 2015
http://www.pfeg.noaa.gov/research/climatemarine/cmfoceanatm/cmfoceanatm2.html	23 June 2015
http://www.climateandweather.net/global-warming/factors-that-influence-climate.html	23 June 2015
http://www.ipcc.ch/publications_and_data/ar4/wg1/en/ch1s1-4-2.html	23 June 2015
http://www.ondrejdanek.cz/rockall/en/rockall.html	
http://www.noc.soton.ac.uk/obe/PROJECTS/EEL/weather.php	
http://www.dcenr.gov.ie/NR/rdonlyres/2A154470-B6B1-495B-B268-D5429B747F13/0/1673R002Annexs2Physicalandchemicalaspects_final.pdf	16 June 2011
http://en.wikipedia.org/wiki/Last_Glacial_Maximum	
http://en.wikipedia.org/wiki/Late_Glacial_Maximum	
http://en.wikipedia.org/wiki/Younger_Dryas	
http://en.wikipedia.org/wiki/File:Golfstream.jpg	
http://oceancurrents.rsmas.miami.edu/atlantic/north-atlantic-drift.htm	24 October 2014
http://en.wikipedia.org/wiki/Dansgaard-Oeschger_event#cite_note-2	
http://en.wikipedia.org/wiki/Ireland#Geograph	
http://en.wikipedia.org/wiki/Oceanic_climate	
http://en.wikipedia.org/wiki/Republic_of_Ireland	
http://en.wikipedia.org/wiki/Ireland	
http://en.wikipedia.org/wiki/County_Kerry	

http://en.wikipedia.org/wiki/Climate_of_south-west_England	
http://en.wikipedia.org/wiki/Geography_of_Cornwall#Climate	
http://en.wikipedia.org/wiki/Devon	
http://en.wikipedia.org/wiki/Wales	
http://en.wikipedia.org/wiki/Faroe_Islands	
http://en.wikipedia.org/wiki/8.2_kiloyear_event	
http://en.wikipedia.org/wiki/List_of_prehistoric_lakes	
http://www.abc.net.au/science/articles/2013/07/04/3796096.htm	19 September 2014
http://en.wikipedia.org/wiki/Deluge_(prehistoric) / https://en.wikipedia.org/wiki/Outburst_flood	18 April 2018
http://en.wikipedia.org/wiki/Altai_flood	29 January 2015
http://en.wikipedia.org/wiki/Fennoscandia	19 September 2014
https://en.wikipedia.org/wiki/Moment_magnitude_scale	
http://en.wikipedia.org/wiki/2011_T%C5%8Dhoku_earthquake_and_tsunami	19 September 2014
http://en.wikipedia.org/wiki/1960_Valdivia_earthquake	27 October 2014
http://en.wikipedia.org/wiki/1815_eruption_of_Mount_Tambora	19 September 2014
http://en.wikipedia.org/wiki/Year_Without_a_Summer	19 September 2014
https://en.wikipedia.org/wiki/List_of_islands_by_area	05 June 2015
http://gizmodo.com/we-were-totally-wrong-about-whats-happening-inside-eart-1775495644	10 May 2016
http://en.wikipedia.org/wiki/Passive_margin	24 November 2012
http://en.wikipedia.org/wiki/Volcanic_passive_margin	20 September 2014
https://en.wikipedia.org/wiki/File:Globald.png	27 July 2013
http://en.wikipedia.org/wiki/Rockall_Basin	20 September 2014
http://en.wikipedia.org/wiki/Hatton_Basin	20 September 2014
http://en.wikipedia.org/wiki/Earth	08 June 2015
http://en.wikipedia.org/wiki/Lithosphere	08 June 2015
http://en.wikipedia.org/wiki/Asthenosphere	08 June 2015
http://goo.gl/zDDQ23	05 June 2015
http://en.wikipedia.org/wiki/Magma	08 June 2015
https://en.wikipedia.org/wiki/Pluton	17 June 2015
http://en.wikipedia.org/wiki/Diapir	08 June 2015

http://en.wikipedia.org/wiki/Mantle_plume	08 June 2015
http://en.wikipedia.org/wiki/Isostacy	11 June 2015
http://www.eurekalert.org/multimedia/pub/91103.php	11 June 2015
http://en.wikipedia.org/wiki/Xiong%27er_Volcanic_Belt	11 June 2015
https://en.wikipedia.org/wiki/Sea_level_rise	12 June 2015
http://worldoceanreview.com/en/wor-1/coasts/sea-level-rise/	12 June 2015
http://en.wikipedia.org/wiki/Post-glacial_rebound	20 September 2014
http://www.telegraph.co.uk/earth/earthnews/6226537/England-is-sinking-while-Scotland-rises-above-sea-levels-according-to-new-study.html	20 September 2014
http://en.wikipedia.org/wiki/Post-glacial_rebound	25 October 2014
http://en.wikipedia.org/wiki/File:Post-glacial_rebound_in_British_Isles.PNG	25 October 2014
https://en.wikipedia.org/wiki/Atlas_Mountains	03 April 2018
http://en.wikipedia.org/wiki/Middle_Atlas	
http://en.wikipedia.org/wiki/High_Atlas	
http://en.wikipedia.org/wiki/Rockall	
http://en.wikipedia.org/wiki/Hasselwood_Rock	
http://en.wikipedia.org/wiki/Iceland_plume	
http://en.wikipedia.org/wiki/Cadiz .	20 September 2014
https://sites.google.com/site/latinjosephus/antiquities/book-1	03 April 2018
http://en.wikipedia.org/wiki/Herodotus	
http://en.wikipedia.org/wiki/Eratosthenes	
www.sea-distances.org	
http://en.wikipedia.org/wiki/Sacred_promontory	27 December 2014
https://en.wikipedia.org/wiki/Bibliotheca_(Pseudo-Apollodorus)	18 April 2018
http://goo.gl/D0QSpl	20 September 2014
http://en.wikipedia.org/wiki/Pherecydes_of_Athens	20 September 2014
http://www.theoi.com/Text/Apollodorus2.html	20 September 2014
http://goo.gl/Zk2o7K	20 September 2014
http://en.wikipedia.org/wiki/Strabo	25 October 2014
https://en.wikipedia.org/wiki/Milesians_(Irish)	

https://en.wikipedia.org/wiki/Colonies_in_antiquity#Greek_colonies	3 April 2018
https://en.wikipedia.org/wiki/Miletus	3 April 2018
http://www.exclassics.com	20 September 2014
http://www.exclassics.com/ceitinn/forintro.htm	20 September 2014
http://en.wikipedia.org/wiki/Goidel_Glas	20 September 2014
http://en.wikipedia.org/wiki/Holinshed's_Chronicles	
http://english.nsms.ox.ac.uk/holinshed/	09 April 2018.
http://goo.gl/nHC07S	09 April 2018
http://en.wikipedia.org/wiki/Raphael_Holinshed	
http://goo.gl/OtCSgh	09 April 2018.
http://english.nsms.ox.ac.uk/holinshed/texts.php?text1=1587_0310	09 April 2018
http://en.wikipedia.org/wiki/Anno_Mundi	
http://en.wikipedia.org/wiki/Éire	09 April 2018
http://en.wikipedia.org/wiki/Goidelic_substrate_hypothesis	29 January 2015
https://en.wikipedia.org/wiki/Circumstantial_evidence	15 March 2016
https://en.wikipedia.org/wiki/Woolly_mammoth .	15 March 2016
https://en.wikipedia.org/wiki/Straight-tusked_elephant	15 March 2016
http://www.bbc.co.uk/nature/20678793	15 March 2016
https://en.wikipedia.org/wiki/Würm_glaciation	04 April 2018
http://www.mineralsireland.ie/Mining+in+Ireland/	21 September 2014
http://www.mineralsireland.ie/MiningInIreland	05 April 2018
http://en.wikipedia.org/wiki/Brass	21 September 2014
http://www.libraryireland.com/SocialHistoryAncientIreland/III-XX-1.php	21 September 2014
http://en.wikipedia.org/wiki/Mining_in_Cornwall_and_Devon	21 September 2014
http://en.wikipedia.org/wiki/Cassiterides	21 September 2014
http://en.wikipedia.org/wiki/Diodorus_Siculus	21 September 2014
http://en.wikipedia.org/wiki/Etruria	21 September 2014
http://en.wikipedia.org/wiki/Upper_Paleolithic	21 September 2014

Endnotes

<u>Note</u>   References in the endnotes to, simply, '*Critias*' or '*Tmaeus*' are references to the to the translation of these works by B Jowett (available at <u>http://www.gutenberg.org/browse/authors/p#a93</u>) while references to '*Perseus Critias*' or '*Perseus Timaeus*' are references to the translations of these works by W.R.M. Lamb (available on the Perseus Tufts website at <u>http://www.perseus.tufts.edu</u>)

1   Muck, Otto Heinrich. *The Secret of Atlantis*. Times Books, 1978.

2   Namely, the author of the blog at <u>http://letseatcakeinwonderland.blogspot.co.za/2011/07/is-atlantis-under-rockall-bank.html</u>, Skender Hushi (<u>http://www.skenderhushi.net/</u>) and the author of the French website <u>http://rockallatlantis.com/francais/cerclespiraleos.html</u>.

3   I do not wish to be seen to be hereby, in any way, belittling or decrying what these other authors have done or achieved.

4   See also the discussion of the ancient Egyptian Edfu texts in Section 2.2. See further the, apparently, similar legend told in ancient Indian epic poems referred to in Alastair Service. *Lost worlds*. Arco Pub., 1981 at p. 195

5   Collina-Girard, Jacques. "L'Atlantide devant le détroit de Gibraltar? Mythe et géologie." / "Atlantis off the Gibraltar Strait? Myth and Geology" *Comptes Rendus de l'Académie des Sciences-Series IIA-Earth and Planetary Science* 333.4 (2001): 233-240 and Gutscher, Marc-André. "Destruction of Atlantis by a great earthquake and tsunami? A geological analysis of the Spartel Bank hypothesis." *Geology* 33.8 (2005): 685-688.

6   See, for example: Vitaliano, D. B., Geomythology. *Journal of the Folklore Institute*, Vol. 5, No. 1 (June 1968), Nunn, P. D. "On the convergence of myth and reality: examples from the Pacific Islands." *The Geographical Journal* 167.2 (2001): 125-138., Nunn, P. D., et al. "Vanished islands in Vanuatu: new research and a preliminary geohazard assessment." *Journal of the Royal Society of New Zealand* 36.1 (2006): 37-50. and Cashman, K. V. and S. J. Cronin. "Welcoming a monster to the world: Myths, oral tradition, and modern societal response to volcanic disasters." *Journal of Volcanology and Geothermal Research* 176.3 (2008): 407-418.).

See also, by way of further examples,a 2001 article was published investigating the possible geological origins for the ancient Greek Delphic oracle (de Boer, JZ, JR. Hale, and J. Chanton. "New evidence for the geological origins of the ancient Delphic oracle (Greece)." *Geology* 29.8 (2001): 707-710.) and a 2014 study conducted to attempt to ascertain whether the ancient Greek myth of Jason and the Golden Fleece was based on actual events. (See: Okrostsvaridze, A., N. Gagnidze, and K. Akimidze. "A modern field investigation of the mythical "gold sands" of the ancient Colchis Kingdom and "Golden Fleece" phenomena." *Quaternary International* (2014). See also: Okrostsvaridze, A. and D. Bluashvili. "Mythical "Gold Sands" of Svaneti (Greater Caucasus, Georgia): Geological Reality and Gold Mining Artefacts." *Bull. Georg. Natl. Acad. Sci* 4.2 (2010).)

7   http://www.edge.org/conversation/a-philosophy-of-physics – dated 30 May 2012 (Accessed on 22 October 2014)

8   Babcock, W. H. *Legendary Islands of the Atlantic: A Study in Medieval Geography*. No. 8. American geographical society, 1922.

9   Roller, Duane W. *Through the pillars of Herakles: Greco-Roman exploration of the Atlantic*. Taylor & Francis, 2006.

10   At p. xvi

11   A balance of probabilities is also referred to as a 'preponderance of evidence'. In essence it means that, having considered all the evidence, there is a greater than 50% chance that the proposition is correct, i.e. the proposition advanced is more probable than not.

12   That said one should take cognisance of the fact that there were very early civilisations which went beyond the general perception of 'stone age' societies. I think of, in particular the Gobekli Tepe and Natufian civilisations. See, for example, in this regard https://en.wikipedia.org/wiki/Gobekli_Tepe, Schmidt, Klaus. *"Göbekli Tepe–the Stone Age Sanctuaries. New results of ongoing excavations with a special focus on sculptures and high reliefs."* Documenta Praehistorica 37 (2010): 239-256., Curry, Andrew, *"Seeking the Roots of Ritual"* Science 319(2008): 278-280, Peters, Joris, and Klaus Schmidt. "Animals in the symbolic world of Pre-Pottery Neolithic Göbekli Tepe, south-eastern Turkey: a preliminary assessment." *Anthropozoologica* 39.1 (2004): 179-218 and https://en.wikipedia.org/wiki/Natufian_culture.

13   Even if what Plato described of the civilisation is not simply a 'vehicle' to make a point it is probable (given the nature of people and age-old tendencies when it comes to inter-group relations and political interests) that many of the descriptions of the population, wealth and power of the people would have been 'massaged' (if not significantly exaggerated) to convey the 'appropriate' perception. This being so (and even accepting that Plato's descriptions of the civilisation were intended to convey what was thought to have existed) there must be significant questions about the veracity of what was stated.

14   See Perseus Tufts translation at par. 113a

15   Clarkson, Chris, et al. *"Human occupation of northern Australia by 65,000 years ago."* Nature 547.7663 (2017): 306.

16   See also, alternatively, Bowler, Jim M., et al. *"Pleistocene human remains from Australia: a living site and human cremation from Lake Mungo, western New South Wales."* World archaeology 2.1 (1970): 39-60.

17   It would appear that I am supported in this view by the authors Elizabeth Wayland Barber and Paul T. Barber in their work *When they severed earth from sky: how the human mind shapes myth*. (Princeton University Press, 2006.) of which I only became aware after having written this section.

18   It is believed that agriculture first started around this time and in response to climate changes.

19   Karmin, Monika, et al. "A recent bottleneck of Y chromosome diversity coincides with a global change in culture." *Genome Research* (2015).

20   "A bottleneck is a period of time when the population size decrease significantly and then recovers. Bottlenecks can be explained by major events such as migration, environmental events, disease or war. Because there are fewer people left after a bottleneck, the population›s offspring have fewer genetic differences … " University of Texas Health Science Center at Houston. «New population genetics model could explain Finn, European genetic differences.» <www.sciencedaily.com/releases/2015/05/150511095358.htm>

21   ScienceDaily. ScienceDaily, 11 May 2015. <www.sciencedaily.com/releases/2015/05/150511095358.htm>

22   University of Texas Health Science Center at Houston. "New population genetics model could explain Finn, European genetic differences." ScienceDaily. ScienceDaily, 11 May 2015. <www.sciencedaily.com/releases/2015/05/150511095358.htm>.

23   See also Xiaoming Liu, Yun-Xin Fu. Exploring population size changes using SNP frequency spectra. *Nature Genetics*, 2015; 47 (5): 555 DOI: 10.1038/ng.3254

24   McMichael, Anthony J. "Insights from past millennia into climatic impacts on human health and survival." *Proceedings of the National Academy of Sciences* 109.13 (2012): 4730-4737.

25   As an aside, in the Eddic poem *Voluspo* describing the end of the world the «Volva,» or wise-woman, who is the narrator of the story, portrays a time when
"I saw there wading | through rivers wild
Treacherous men | and murderers too
And workers of ill | with the wives of men;
There Nithhogg sucked | the blood of the slain
And the wolf tore men; | ….."
and
"Brothers shall fight | and fell each other,
And sisters› sons | shall kinship stain;
Hard is it on earth, | with mighty whoredom;
Axe-time, sword-time, | shields are sundered,
Wind-time, wolf-time, | ere the world falls;
Nor ever shall men | each other spare."

26   The word literally means "the recitation", (it is also referred to, in English, as the Koran). It is the central religious text of Islam. See http://en.wikipedia.org/wiki/Quran (Accessed 26 October 2014)

27   http://en.wikipedia.org/wiki/Hafiz_%28Quran%29 (Accessed 26 October 2014)

28   http://en.wikipedia.org/wiki/Quran (Accessed 26 October 2014)

29   Nunn, P, *The Edge of Memory: The Geology of Folk Tales and Climate Change*, Bloomsbury Publishing Plc, 2018, Chapter One.

30   Masse, W. Bruce, et al. "Exploring the nature of myth and its role in science." *Geological Society, London, Special Publications* 273.1 (2007): 9-28 at p. 18, state that "Mythology is a function of the oral transmission of linguistically

encoded data" and that "Myths are not just 'silly cultural fiction', they are carriers of information once deemed extremely important"

31   See also Cashman, Katharine V., and Shane J. Cronin. "Welcoming a monster to the world: Myths, oral tradition, and modern societal response to volcanic disasters." *Journal of Volcanology and Geothermal Research* 176.3 (2008): 407-418.

32   For some useful definitions of myths and geomythology see Masse, W. Bruce, et al. "Exploring the nature of myth and its role in science." *Geological Society, London, Special Publications* 273.1 (2007): 9-28 at p 17 and Vitaliano, Dorothy B. "Geomythology: geological origins of myths and legends." *Geological Society, London, Special Publications* 273.1 (2007): 1-7. at pp 2, 3

33   Vitaliano, D. B., (1968). Geomythology. *Journal of the Folklore Institute*, Vol. 5, No. 1 (June 1968)

34   See also https://en.wikipedia.org/wiki/Geomythology (Accessed 27 December 2014)

35   Echo-Hawk, Roger C. «Ancient history in the New World: integrating oral traditions and the archaeological record in deep time.» *American Antiquity* 65.2 (2000): 267-290.

36   Bryant, E., G. Walsh, and D. Abbott. "Cosmogenic mega-tsunami in the Australia region: are they supported by Aboriginal and Maori legends?." *Geological Society, London, Special Publications* 273.1 (2007): 203-214.

37   Rappenglück, B., et al. "The fall of Phaethon: a Greco-Roman geomyth preserves the memory of a meteorite impact in Bavaria (south-east Germany)." *Antiquity* 84.324 (2010): 428-439.

38   Vitaliano, D. B. "Geomythology: geological origins of myths and legends." *Geological Society, London, Special Publications* 273.1 (2007): 1-7. at pp 2, 3

39   Zdanowicz, C. M., G. A. Zielinski, and M. S. Germani. "Mount Mazama eruption: Calendrical age verified and atmospheric impact assessed." *Geology* 27.7 (1999): 621-624.

40   See http://en.wikipedia.org/wiki/Mount_Mazama (Accessed on 24 May 2015).

41   Deur, Douglas. «A most sacred place: The significance of Crater Lake among the Indians of Southern Oregon.» *Oregon Historical Quarterly* 103.1 (2002): 18-49.

42   *Yidyam* (Lake Eacham) and *Barany* (Lake Barrine)

43   See http://en.wikipedia.org/wiki/Lake_Euramo (Accessed 26 October 2014)

44   See also Dixon, R. M. W. "The Dyirbal language of North Queensland." *Cambridge Studies in Linguistics, Cambridge University Press* (1972). and Dixon, R.M.W., "Origin legends and linguistic relationships" *Oceania* 67, 127-139 (1996). http://www.thefreelibrary.com/ Origin+legends+and+linguistic+relationships.-a019175999 (Accessed 26 October 2014)

45   Nunn, P. D. and N. J. Reid *"Aboriginal Memories of Inundation of the Australian Coast Dating from More than 7000 Years Ago" Australian Geographer*, *DOI: 10.1080/00049182.2015.1077539*, 2015

46   See http://en.wikipedia.org/wiki/Prose_Edda (Accessed on 26 October 2014)

47   The Edda comprises both the Prose Edda and the Poetic Edda. Together these comprise the major store of pagan Scandinavian mythology. The former is assumed to have been written, or at least compiled, by the Icelandic scholar and historian Snorri Sturluson around the year 1220. The Poetic Edda is a collection of anonymous poetry from earlier traditional sources and must pre-date the Prose Edda as the latter cites various poems collected in the Poetic Edda as sources.

48   Mörner, N-A. "The Fenris Wolf in the Nordic Asa creed in the light of palaeoseismics." *Geological Society, London, Special Publications* 273.1 (2007): 117-119. at p. 117

49   At pp 118/9

50   Veski, S., et al. "The physical and social effects of the Kaali meteorite impact—a review." *Comet/asteroid impacts and human society*. Springer Berlin Heidelberg, 2007. 265-275.

51   Raukas, A. and W. Stankowski. "On the age of the Kaali craters, Island of Saaremaa, Estonia." *Baltica* 24.1 (2011): 37-44.

52   Also referred to as the "Aeta".

53   The Ayta often refer to themselves as the Katutubo which may be translated as "the ones from the land". See Shimizu, H. "Past, Present and Future of the Pinatubo Aetas: At the Crossroads of Socio-Cultural Disintegration." *After the Eruption: Pinatubo Aetas at the Crisis of their Survival* (1992): 1-49.

54   The Aytas, or their predecessors, probably arrived in the Philippines at about 18 000 BP. See A prehistoric lahar-dammed lake and eruption of Mount Pinatubo described in a Philippine aborigine legend, infra at fn. 596, p 435

55   Rodolfo, K. S., and J. V. Umbal. "A prehistoric lahar-dammed lake and eruption of Mount Pinatubo described in a Philippine aborigine legend." *Journal of Volcanology and Geothermal Research* 176.3 (2008): 432-437.

56   Newhall, C. G., and R. Punongbayan, eds. *Fire and mud: eruptions and lahars of Mount Pinatubo, Philippines*. Quezon City: Philippine Institute of Volcanology and Seismology, 1996 as cited in Rodolfo, K. S., and J. V. Umbal. "A prehistoric lahar-dammed lake and eruption of Mount Pinatubo described in a Philippine aborigine legend." *Journal of Volcanology and Geothermal Research* 176.3 (2008): 432-437.

57   The Maraunot and Buag eruptions

58   http://en.wikipedia.org/wiki/Solon (Accessed on 25 October 2014)

59   *Perseus Timaeus* at par. 21e

60   Sais was the Greek name for the city but the ancient Egyptian name for the city was Sa or Zau. It is now known as Sa el-Hagar See http://en.wikipedia. org/wiki/Sais,_Egypt (Accessed on 25 October 2014)

61   *Perseus Timaeus* at par. 21E, Lesko, Barbara S. *The great goddesses of Egypt*. University of Oklahoma Press, 1999, p. 45, Budge, Ernest Alfred Tompson Wallis. *The Gods of the Egyptians, Or Studies in Egyptian Mythology, by EA Wallis Budge,...* Methuen, 1904. at p. 20

62   Her name was also spelled also spelled Nit, Net, or Neit. See http:// en.wikipedia.org/wiki/Neith (Accessed 25 October 2014)

63   There are suggestions that Athena, the patron goddess of Athens (who, reputedly, originally came from Libya) may be a Greek form of Neith.

64   Lesko, B. S. *The great goddesses of Egypt*. University of Oklahoma Press, 1999, p. 45. It is even possible that Neith was a Libyan goddess adopted by the Egyptians in pre-dynastic times – see Budge, Ernest Alfred Tompson Wallis. *The Gods of the Egyptians, Or Studies in Egyptian Mythology, by EA Wallis Budge,* Methuen, 1904. at p. 275

65   Lesko, B. S. *The great goddesses of Egypt*. University of Oklahoma Press, 1999. and http://en.wikipedia.org/wiki/Neith (Accessed 25 October 2014)

66   According to Plutarch, in his life of Solon, the priest was *"Sonchis the Saite, the most learned of all the priests"*. See also http://en.wikipedia.org/wiki/Sonchis_of_Sais (Accessed 25 October 2014)

67   *Timaeus* .

Note   References in these endnotes to, simply, '*Critias*' or '*Tmaeus*' are references to the to the translation of these works by B Jowett (available at http://www.gutenberg.org/browse/authors/p#a93) while references to '*Perseus Critias*' or '*Perseus Timaeus*' are references to the translations of these works by W.R.M. Lamb (available on the Perseus Tufts website at http://www.perseus.tufts.edu)

68   *Critias*.

Note   References in these endnotes to, simply, '*Critias*' or '*Tmaeus*' are references to the to the translation of these works by B Jowett (available at http://www.gutenberg.org/browse/authors/p#a93) while references to '*Perseus Critias*' or '*Perseus Timaeus*' are references to the translations of these works by W.R.M. Lamb (available on the Perseus Tufts website at http://www.perseus.tufts.edu)

69   See http://www.iep.utm.edu/plato/ (Accessed 26 October 2014), http://www.ancient.eu.com/plato/ (Accessed 26 October 2014) and http://en.wikipedia.org/wiki/Plato (Accessed 26 October 2014)

70   Strabo (17.29)

71   I also draw support for this contention from a recent article (Nunn, P. D. & N. J. Reid [2015]: Aboriginal Memories of Inundation of the Australian Coast Dating from More than 7000 Years Ago, Australian Geographer, in which the authors conclude that the are Australian Aboriginal stories and myths which might date as far back as 13 070 cal. years BP. (i.e. approximately 11 700 BCE)

72   McClain, Brett, 2011, Cosmogony (Late to Ptolemaic and Roman Periods). In Jacco Dieleman, Willeke Wendrich (eds.), UCLA Encyclopedia of Egyptology, Los Angeles.

73   See also Hollis, Susan Tower. "Otiose Deities and the Ancient Egyptian Pantheon." *Journal of the American Research Center in Egypt* 35 (1998): 61-72.

74   Jelinkova, ⊠. AE. «The Shebtiw in the temple at Edfu.» *Zeitschrift für Ägyptische Sprache und Altertumskunde* 87.1-2 (1962): 41-54

75   https://en.wikipedia.org/wiki/Heryshaf Accessed on 08 July 2018.

76   Reymond, Eve AE. *The mythical origin of the Egyptian temple*. Manchester University Press, 1969.

In referring to this work I acknowledge that Reymond probably did not put the same
interpretation on her findings as I propose to do however my proposition is
consistent with her findings

77   See pp. 60, 74, 76 and 101

78   See p. 60

79   See p. 69

80   See pp. 80 and 89

81   See pp. 105 and 108

82   See p. 113. It may also be interesting to note that falling comets or meteors
often have an association with snakes (or dragons) and that it appears to be
becoming more scientifically accepted that there was an impact by a comet
or meteorite at the boundary between the end of the Pleistocene and the
beginning of the Holocene. See Wolbach, Wendy S., et al. «Extraordinary
biomass-burning episode and impact winter triggered by the Younger
Dryas cosmic impact ~ 12,800 years ago. 1. Ice cores and glaciers.» *The
Journal of Geology* 126.2 (2018): 165-184 and also Wolbach, Wendy S., et
al. «Extraordinary biomass-burning episode and impact winter triggered by
the Younger Dryas cosmic impact~ 12,800 years ago. 2. Lake, marine, and
terrestrial sediments.» *The Journal of Geology* 126.2 (2018): 185-205.

83   See p. 109

84   See pp. 106 and 112

85   See p. 113/4

86   See pp. 81, 119 and 120

87   See p. 13

88   See p. 109

89   See p. 116

90   Insofar as it may be queried as to why, or how, an Egyptian temple would have
been likely to have any record of an island in the Atlantic it may, in addition
to any other factors, be borne in mind that the Edfu temple complex was
a Ptolemaic complex and that the Ptolemaic dynasty had Greek links. See
https://en.wikipedia.org/wiki/Ptolemaic_dynasty (Accessed on 1 August 2018)

91   There is even a controversy as to whether or not Homer was a single person.

92   See http://en.wikipedia.org/wiki/Homer .

93   The Project Gutenberg EBook of *The Odyssey*, by Homer, translated by Samuel
Butler. [Butler, Samuel. *The Odyssey render into English prose for the use of those
who cannot read the original.* 1922.] (See also, for an alternate translation, The
Project Gutenberg EBook of *The Odyssey* of Homer, by Homer, translated by
William Cowper) [The Odyssey of Homer translated by Cowper, W, London:
J·M·Dent·&·Sons·Ltd and in New York by E·P·Dutton & Co 1820]

94   *The Geography of Strabo* published in Vol. I of the Loeb Classical Library
edition, 1917

95   http://penelope.uchicago.edu/Thayer/E/Roman/Texts/Strabo/1B1*.html
(Accessed 21 September 2014)

96   Who lived from 23CE to 79CE.

97   Plin. Nat. 4.30 - Pliny's Natural History. In thirty-seven books. A translation
       on the basis of that by Dr. Philemon Holland, Ed. 1601. With critical and
       explanatory notes by the Wernerian Club, 1847/8.

98   Plutarch, *Moralia*, Vol XII, 26 *Concerning the Face Which Appears in the Orb of
       the Moon*

99   See https://en.wikipedia.org/wiki/Rockall (Accessed on 11 August 2015)

100   McGrail, Seán. *Ancient Boats in North-West Europe: The archaeology of water
       transport to AD 1500*. Routledge, 2014. at p263

101   See https://en.wikipedia.org/wiki/Olaus_Rudbeck (Accessed on 13 August
       2015.)

102   From *The Odyssey of Homer.* , translated by William Cowper, London:
       Published by J·M·Dent·&·Sons Ltd and in New York by E·P·Dutton & Co,
       1820. (See also The Odyssey with an English Translation by A.T. Murray,
       PH.D. in two volumes. Cambridge, MA., Harvard University Press; London,
       William Heinemann, Ltd. 1919 at http://www.perseus.tufts.edu/hopper/
       text?doc=Perseus%3Atext%3A1999.01.0136%3Abook%3D5%3Acard%-
       3D262 .])

103   Zhirov, Nicolas. *Atlantis: Atlantology: Basic Problems*. University Press of the
       Pacific, 2001. at p. 75

104   Rudbeck, in attempting to determine the position of Ogygia, referred, in
       particular, to the statement, in Book V, line 270, that "*Gladly then did goodly
       Odysseus spread his sail to the breeze; and he sat and guided his raft skilfully with
       the steering-oar, nor did sleep fall upon his eyelids, as he watched the Pleiads, and
       late-setting Bootes, and the Bear, which men also call the Wain, which ever circles
       where it is and watches Orion, and alone has no part in the baths of Ocean. For
       this star Calypso, the beautiful goddess, had bidden him to keep on the left hand as
       he sailed over the sea.*"
       That section has also been translated as "*Beside the helm he sat, steering expert,
       Nor sleep fell ever on his eyes that watch'd Intent the Pleiads, tardy in decline
       Bootes, and the Bear, call'd else the Wain, Which, <u>in his polar prison circling</u>*, [my
       underling] *looks Direct toward Orion, and alone Of these sinks never to the briny
       Deep. That star the lovely Goddess bade him hold Continual on his left through all
       his course.*"

105   De Camp, L. Sprague. *Lost Continents*. Gnome Press, 1954. at p. 178.
       See also https://en.wikipedia.org/wiki/Mecklenburg (Accessed on 13 August
       2015)

106   See https://en.wikipedia.org/wiki/Rockall (Accessed on 13 August 2015.)

107   See http://www.mareano.no/en/maps/mareano_en.html (Accessed on 13
       August 2015.)

108   de Mirabilibus Auscultationibus as published in the Loeb Classical Library
       Cambridge (Mass.) and London, 1936. The text is in the public domain
       and was obtained at http://penelope.uchicago.edu/Thayer/E/Roman/Texts/
       Aristotle/de_Mirabilibus*.html (Accessed on 10 November 2018)

109   See http://bmcr.brynmawr.edu/2009/2009-02-22.html and https://
       en.wikipedia.org/wiki/Paradoxography#cite_note-1 (Accessed on 20
       November 2019)

110   de Mirabilibus Auscultationibus at par. 84 – See end note above

111   Aristotle, W. D., and J. A. Ross. "Works Translated Into English Under the Editorship of WD Ross." (1908) containing " De Mirabilibus auscultationibus." translated by Dowdall, D at par. 84.

112   https://en.wikipedia.org/wiki/Proclus (Accessed on 10 November 2018)

113   *Commentaries of Proclus on the Timaeus of Plato, in five books; containing a treasury of Pythagoric and Platonic physiology* by Proclus Diadochus, the Platonic Successor Translated from the Greek by Thomas Taylor London: 1820

114   https://maps.ngdc.noaa.gov/viewers/bathymetry/ (Accessed on 14 November 2018)

115   A nymph in Greek and in Latin mythology is a minor female nature deity typically associated with a particular location or landform. See http://en.wikipedia.org/wiki/Nymph (Accessed on 19 September 2014)

116   See http://www.theoi.com/Titan/Hesperides.html (Accessed on 19 September 2014)

117   See http://en.wikipedia.org/wiki/Hesperides (Accessed on 19 September 2014)

118   See http://www.perseus.tufts.edu/Herakles/apples.html (Accessed on 19 September 2014)

119   See http://penelope.uchicago.edu/Thayer/E/Roman/Texts/Diodorus_Siculus/4B*.html (Accessed on 19 September 2014)

120   See http://www.theoi.com/Titan/Hesperides.html (Accessed on 19 September 2014)

121   See http://en.wikipedia.org/wiki/Hesperides (Accessed on 19 September 2014)

122   See http://www.theoi.com/Titan/Hesperides.html (Accessed on 19 September 2014) I shall refer to the island of Erytheia again in section 8.

123   Referred to below.

124   The Shorter Oxford English Dictionary on historical principles, (2002). (Vol. 1 A-M). Oxford University Press.

125   See https://translate.google.co.za/?hl=en&tab=wT#el/en/%CE%BC%-CE%B7%CE%BB%CE%BF%CF%82 .. and https://translate.google.co.za/?hl=en&tab=wT#el/en/%CE%95%E1%BD%94%CE%BC%CE%-B7%CE%BB%CE%BF%CF%82. (Accessed on 2 February 2015)

126   On the other hand Zhirov (Zhirov, N.. *Atlantis: Atlantology: Basic Problems.* University Press of the Pacific, 2001. at p. 28) states that it translates as "owner of fine sheep".

127   See section 8.3.3

128   See https://en.wikipedia.org/wiki/Hecataeus_of_Abdera

129   He wrote the Library of History between 60 and 30 BCE.

130   See https://en.wikipedia.org/wiki/Pseudo-Scymnus (Accessed on 7 September 2018)

131   Pseudo Scymnus', or Pausanias of Damascus', *Circuit of the Earth* (unedited translation by Brady Kiesling from the Muller text *(Geographi Graeci Minores)* of 1857) accessed at https://topostext.org/work/130 (Accessed on 7 September 2018)

132   Other variants of the name are Hy Brazil, Hy Breasil, Hy Breasail, Hy Breasal, Hy Brasil. See Wikipedia http://en.wikipedia.org/wiki/Brazil

(mythical_island) Also Brasil, Bersil, Brazir, O'Brazil, O'Brassil, Breasail See also Babcock, W. H.. *Legendary Islands of the Atlantic: A Study in Medieval Geography*. No. 8. American geographical society, 1922.

133   Strictly speaking a 'portolan chart'.

134   Including a Catalan chart of about 1480.

135   See Wikipedia http://en.wikipedia.org/wiki/Brazil_(mythical_island)

136   See also Vinson, D. A. "The Western Sea: Atlantic History before Columbus." *Northern Mariner* 10.3 (2000): 1-14.

137   See Wikipedia http://en.wikipedia.org/wiki/Brazil_(mythical_island)

138   Babcock, W. H.. *Legendary Islands of the Atlantic: A Study in Medieval Geography*. No. 8. American geographical society, 1922.

139   See http://en.wikipedia.org/wiki/Avalon, (Accessed 9 June 2014)

140   See http://en.wikipedia.org/wiki/Avalon, (Accessed 9 June 2014)

141   See also http://en.wikipedia.org/wiki/Emain_Ablach, (Accessed 9 June 2014) for a reference to another Isle of Apples in Irish mythology.

142   See http://www.sacred-texts.com/neu/eng/vm/vmeng.htm (Accessed 9 June 2014)

143   W. Y. Evans Wentz, in his book "*The Fairy-Faith In Celtic Countries*" [Wentz, W. Y. "Evans. The Fairy-Faith in Celtic Countries." (1911)], unambiguously links Avalon with the 'Otherworld' of Celtic beliefs, as does the author of Jones' Celtic Encyclopedia. [See http://www.maryjones.us/jce/otherworld.html, (Accessed 9 June 2014)] The Otherworld was, for Celts, not a separate "heaven", as envisaged by the Abrahamic faiths, but existed nearby but just out of reach. (See Monaghan, Patricia. *The encyclopedia of Celtic mythology and folklore*. Infobase Publishing, 2009.)

144   I shall for the most part use the translations by Benjamin Jowett. These may be found in Jowett, B. "The dialogues of Plato, translated into English with analyses and introductions (Vol. 3)." (1892) Oxford Clarendon Press. These translations, both the *Timaeus* and the *Critias,* are downloadable individually at Project Gutenberg at http://www.gutenberg.org/browse/authors/p#a93. I shall refer, in the footnotes, to these translations simply as, respectively, the *Timaeus* and the *Critias.*

145   I shall also make reference to the translation by W.R.M. Lamb – Plato in Twelve Volumes, Vol. 9 translated by W.R.M. Lamb. Cambridge, MA, Harvard University Press; London, William Heinemann Ltd. 1925. These translation may be viewed at http://www.perseus.tufts.edu . I shall refer, in the footnotes, to these translations as the *Perseus Timaeus* or, as the case might be, the *Perseus Critias.*

146   http://en.wikipedia.org/wiki/Plato (Accessed 26 October 2014)

147   Taylor, A. E. "Plato, the Man and His Work. 7th--ed." *London: Methuen* (1960).

148   See also http://en.wikipedia.org/wiki/Plato (Accessed on 3 February 2015)

149   See also The Life of Plato by Diogenes Laertius in Burges, G "The Works of Plato, a new and literal version. Vol. IV containing the doubtful works, Lives of Plato, by Diogenes Laertius, Hesychius and Olympiodorus ; Introductions

to his doctrines, by Alcinous and Albinus. Published by H.G. Bohn, (1854) p. 175 *et seq.*

150   See Gibson, T. "Epilogue to Plato: The bias of literacy." *Proceedings of the Media Ecology Association.* Vol. 6. 2005 and, as referred to by Gibson, Havelock, E. A. *Preface to Plato.* Vol. 1. Harvard University Press, 2009.

151   See previous footnote.

152   Gibson, T. "Epilogue to Plato: The bias of literacy." *Proceedings of the Media Ecology Association.* Vol. 6. 2005 in the 'Conclusion'.

153   Taylor, A. E. "Plato, the Man and His Work. 7[Th] ed." *London: Methuen* (1960) at p. 24

154   Taylor, A. E. "Plato, the Man and His Work. 7[Th] ed." *London: Methuen* (1960) at p. 23

155   See, in particular, his *Phaedrus* and his *Seventh Letter* .

156   Plato. *Letter No. 7* at par. 344c, from "Plato. Plato in Twelve Volumes", Vol. 7 translated by R.G. Bury. Cambridge, MA, Harvard University Press; London, William Heinemann Ltd. 1966. obtained from perseus.tufts.edu - http://goo.gl/Ql1LUp (Accessed on 6 February 2015)

157   The sentence continues " … *lest thereby he may possibly cast them as a prey to the envy and stupidity of the public.*"

158   Plato, *Letter No. 7* at par. 344c, from "Plato. Plato in Twelve Volumes", Vol. 7 translated by R.G. Bury. Cambridge, MA, Harvard University Press; London, William Heinemann Ltd. 1966. obtained from perseus.tufts.edu - http://goo.gl/AUatI1 (Accessed on 6 February 2015)

159   Taylor, A. E. "Plato, the Man and His Work. 7th--ed." *London: Methuen* (1960) at page 23.

160   Plato, *Phaedrus* at par. 276d from "Plato. Plato in Twelve Volumes", Vol. 7 translated by R.G. Bury. Cambridge, MA, Harvard University Press; London, William Heinemann Ltd. 1966. obtained from perseus.tufts.edu - http://goo.gl/IEyITN (Accessed on 6 February 2015)

161   Taylor, A. E. "Plato, the Man and His Work. 7th--ed." *London: Methuen* (1960) at p. 2.

162   *Critias*

163   See also *Perseus Critias* at paragraph 113b.

164   "The **Republic ...** is a Socratic dialogue, written by Plato around 380 BCE, concerning the definition of justice (δικαιοσύνη), the order and character of the just city-state and the just man ..." See https://en.wikipedia.org/wiki/Republic_%28Plato%29

165   Taylor, A. E. "Plato, the Man and His Work. 7th--ed." *London: Methuen* (1960) at page 436

166   Heracles, Herakles and Hercules are synonyms. The latter form, Hercules, is the Roman spelling of the name. See also the Perseus Tufts / Lamb translation at par. 108e

167   See also Perseus Tufts / Lamb translation at par. 108e

168   *Critias*. See also Perseus Tufts / Lamb translation at par. 108e.

169   The term the 'Pillars of Heracles' is generally accepted as being a reference to the straits of Gibraltar.

170   *Timaeus.* See also Perseus Tufts / Lamb translation at par's. 25e and 25de.

171   Plat. Laws 3.677 at http://www.perseus.tufts.edu/hopper/text?doc=Perseus%-3Atext%3A1999.01.0165%3Abook%3D3%3Apage%3D677

172   Plato, and Alfred Edward Taylor. *The Laws of Plato. Translated Into English by AE Taylor.[With Introduction and Notes.].* JM Dent & Sons, 1934.at p. 56

173   Plato lived between, about, 428 and 348 BCE and said that the sinking had taken place about 9000 years before that.

174   *The History of Herodotus* by Herodotus, Book 1, Clio Translated by George Rawlinson (See also, for an alternate translation, *The History of Herodotus*, by Herodotus. Translated into English by G. C. Macaulay.)

175   The Erythraean sea is, it is generally accepted, what is now referred to as the Indian Ocean.

176   I understand from this that Herodotus recognised that all the oceans are one body of water and that the divisions into the Atlantic and other oceans are merely conventions.

177   See also Plutarch's reference to the '*great mainland, by which the great ocean is encircled* ... ", referred to in paragraph 3.3.2, in relation to the island of Ogygia.

178   The straits of Gibraltar.

179   There is no agreement as to what, if any, was the standard length of a stade in modern measurements. All stades were 600 feet long but the measurement of a foot varied, from 262 mm to 349 mm. The itinerary stade (which was used for measuring journeys) is 157 metres long. (See http://en.wikipedia.org/wiki/Stadion_%28unit_of_length%29).

180   *Critias.* See also Perseus Tufts / Lamb translation at par. 113c

181   *Critias.* See also Perseus Tufts / Lamb translation at par. 113e

182   *Critias.* See also Perseus Tufts / Lamb translation at par. 114e

183   *Critias.* See also Perseus Tufts / Lamb translation at par. 114e to 115b

184   *Critias.* See also Perseus Tufts / Lamb translation at par. 117a

185   i.e. 471 kilometres – based upon the itinerary stade.

186   i.e. 314 kilometres – based upon the itinerary stade.

187   *Critias.* See also Perseus Tufts / Lamb translation at par. 118a

188   Perseus Tufts / Lamb translation at par. 118b

189   *Critias.* See also Perseus Tufts / Lamb translation at par. 118c

190   *Critias.* See also Perseus Tufts / Lamb translation at par. 118e

191   *Critias.* See also Perseus Tufts / Lamb translation at par. 114e

192   Perseus Tufts / Lamb translation at par. 116c

193   *Critias*

194   According to the *Perseus Critias* (par. 118a) translation "this plain had a level surface and was as a whole rectangular in shape, being 3000 stades long on either side and 2000 stades wide at its centre".

195   Perseus Tufts / Lam*b Critias*, par. 118c - d

196   *Critias,*

197   *Critias*

198   *Perseus Critias*, par. 118d

199   Whether this could have been zones of sea, as distinct from water, may however be called into question by the fact that it is later mentioned that later rulers *"bored a canal of three hundred feet in width and one hundred feet in depth and fifty stadia in length, which they carried through to the outermost zone, making a passage from the sea up to this, which became a harbour, and leaving an opening sufficient to enable the largest vessels to find ingress"*.

200   *Critias*

201   Perseus Tufts / Lamb *Critias*, par's 113c and 113d.

202   *Perseus Critias* at par's 113e to 114b

203   Timaeus 24e, Lamb translation - Plato in Twelve Volumes, Vol. 9 translated by W.R.M. Lamb

204   Critias 108e, Lamb translation - Plato in Twelve Volumes, Vol. 9 translated by W.R.M. Lamb

205   Lamb translation - Plato in Twelve Volumes, Vol. 9 translated by W.R.M. Lamb

206   Lamb translation - Plato in Twelve Volumes, Vol. 9 translated by W.R.M. Lamb

207   Lamb translation - Plato in Twelve Volumes, Vol. 9 translated by W.R.M. Lamb

208   Now known as the Rioni or Rion River, it is the main river of western Georgia originating in the Caucasus Mountains and flows, west, into the Black Sea.

209   Plato. Plato in Twelve Volumes, Vol. 1 translated by Harold North Fowler; Introduction by W.R.M. Lamb. Cambridge, MA, Harvard University Press; London, William Heinemann Ltd. 1966 at http://www.perseus.tufts.edu/hopper/text?doc=Perseus:text:1999.01.0170:text=Phaedo:section=109a

210   Plat. Tim. 25a, Plato, Plato in Twelve Volumes, Vol. 9 translated by W.R.M. Lamb

211   Perseus Tufts / Lamb *Timaeus*, par. 25b

212   Or, in the Jowett translation, within the Pillars of Heracles.

213   Extracted from Google Maps

214   Hom. Il. 14.193 from Homer. The Iliad with an English Translation by A.T. Murray, Ph.D. in two volumes. Cambridge, MA., Harvard University Press; London, William Heinemann, Ltd. 1924 at http://www.perseus.tufts.edu/hopper/text?doc=Perseus%3Atext%3A1999.01.0134%3Abook%3D14%-3Acard%3D193

215   Hom. Il. 21.161

216   http://www.theoi.com/Kosmos/Okeanos.html

217   https://en.wikipedia.org/wiki/Hecataeus_of_Miletus

218   Shefton, Brian Benjamin, and Kathryn Lomas, eds. *Greek identity in the Western Mediterranean: papers in honour of Brian Shefton*. Vol. 246. Brill, 2004. at p. 301.

219   Pind. I. 4 from Odes. Pindar. Diane Arnson Svarlien. 1990 at http://www.perseus.tufts.edu/hopper/text?doc=Perseus%3Atext%3A1999.01.0162%-3Abook%3DI.%3Apoem%3D4

220   Pind. N. 3 (Nemean 3, For Aristocleides of Aegina Pancratium) from Odes. Pindar. Diane Arnson Svarlien. 1990 at http://www.perseus.tufts.edu/hopper/

text?doc=Perseus%3Atext%3A1999.01.0162%3Abook%3DN.%3Apoem%-3D3

221   Pind. O. 2.70 from Odes. Pindar. Diane Arnson Svarlien. 1990 at http://www.perseus.tufts.edu/hopper/text?doc=Perseus%3Atext%3A1999.01.0162%-3Abook%3DO.%3Apoem%3D2

222   Strab. 3.5.5 from The Geography of Strabo. Literally translated, with notes, in three volumes. London. George Bell & Sons. 1903 at http://www.perseus.tufts.edu/hopper/text?doc=Perseus%3Atext%3A1999.01.0239%3Abook%-3D3%3Achapter%3D5%3Asection%3D5

223   I, however, propose in Section 10 that the Gades/Gadeira referred to by Plato was Ireland.

224   A document, or type of log, listing ports and coastal landmarks, in order and with approximate intervening distances, used by ancient sailors for navigation.

225   Hdt. 4.8.2

226   Presumably the Mediterranean.

227   Hdt. 1.202.4 / 1.203.1 from Herodotus, with an English translation by A. D. Godley. Cambridge. Harvard University Press. 1920 at http://www.perseus.tufts.edu/hopper/text?doc=Perseus%3Atext%3A1999.01.0126%3Abook%-3D1%3Achapter%3D202%3Asection%3D4

228   Hdt. 4.152 Herodotus, with an English translation by A. D. Godley. Cambridge. Harvard University Press. 1920. at http://www.perseus.tufts.edu/hopper/text?doc=Perseus%3Atext%3A1999.01.0126%3Abook%3D4%-3Achapter%3D152

229   https://en.wikipedia.org/wiki/Colaeus

230   https://en.wikipedia.org/wiki/Tartessos

231   The Black Sea

232   Euripides. Euripides, with an English translation by David Kovacs. Cambridge. Harvard University Press at http://www.perseus.tufts.edu/hopper/text?doc=Perseus%3Atext%3A1999.01.0106%3Acard%3D1

233   Also translated as *"Wide o'er man my realm extends, and proud the name that I, the goddess Cypris, bear, both in heaven's courts and 'mongst all those who dwell within the limits of the sea and the bounds of Atlas ..."* Hippolytus, by Euripides Translated by E. P. Coleridge at http://classics.mit.edu/Euripides/hippolytus.html

234   Hippolytus and The Bacchae, by Euripides, translated by Gilbert Murray at http://www.gutenberg.org/files/8418/8418-h/8418-h.htm

235   Also translated as *"To the apple-bearing shore of the Hesperides, famous singers, would I go my way, there where the lord of the deep-blue mere forbids further passage to sailors, fixing the sacred boundary of the skies, the pillar held up by Atlas. "* at http://www.perseus.tufts.edu/hopper/text?doc=Perseus%3Atext%-3A1999.01.0106%3Acard%3D742

236   https://en.wikipedia.org/wiki/Periplus_of_Pseudo-Scylax

237   Herodotus uses the term Ethiopia to refer to those parts of Africa which as were then known to fall within the habitable world.

238   Pseudo-Skylax's Periplous or Circumnavigation, translation by Brady Kiesling (2015) from the 1878 Greek edition of B. Fabricius from http://topostext.org/work.php?work_id=102

239   Aristot. Rh. 2.10 from Aristotle in 23 Volumes, Vol. 22, translated by J. H. Freese. Aristotle. Cambridge and London. Harvard University Press; William Heinemann Ltd. 1926 at http://www.perseus.tufts.edu/hopper/text?doc=Perseus:text:1999.01.0060:book=2:chapter=10&highlight=pillars

240   Meteorology by Aristotle translated by E. W. Webster at http://classics.mit.edu/Aristotle/meteorology.2.ii.html

241   Ælian's Various History, The Fifth Book at http://penelope.uchicago.edu/aelian/varhist5.xhtml

242   See https://en.wikipedia.org/wiki/Pseudo-Aristotle and https://en.wikipedia.org/wiki/On_the_Universe

243   *De mundo* by Aristotle; tr. by Forster, E. S. and Dobson, J. F. (1914) published by Oxford : The Clarendon Press at https://archive.org/details/demundoarisrich

244   Strabo III.5.5 at http://penelope.uchicago.edu/Thayer/E/Roman/Texts/Polybius/34*.html (Accessed on 16 April 2018)

245   See respective references at https://en.wikipedia.org/wiki/ (Accessed on 16 April 2018)

246   See https://en.wikipedia.org/wiki/Oceanus

247   Plat. Phaedo 112/3 from Plato in Twelve Volumes, Vol. 1 translated by Harold North Fowler; Introduction by W.R.M. Lamb. Cambridge, MA, Harvard University Press; London, William Heinemann Ltd. 1966 at http://www.perseus.tufts.edu/hopper/text?doc=Perseus%3Atext%3A1999.01.0170%-3Atext%3DPhaedo%3Apage%3D112

248   Scholiast of Apollonius Rhodius. 1. 211

249   See also Mangas, Julio, et al., eds. *La Península Ibérica en los autores griegos: de Homero a Platón*. Editorial Complutense, 1998.

250   https://en.wikipedia.org/wiki/Stesichorus

251   iv. 2.22 from *Physical science in the time of Nero; being a translation of the Quaestiones naturales of Seneca* by Seneca tr. by Clarke, J, and Geikie, A, Sir, (1910) London : Macmillan and Co., Limited at https://archive.org/details/physicalsciencei00seneiala

252   Hdt. 2 32-33 "as to the river that ran past the city, Etearchus guessed it to be the Nile; and that is but reasonable."

253   Hdt. 4.8.2 from Herodotus, with an English translation by A. D. Godley. Cambridge. Harvard University Press. 1920 at http://www.perseus.tufts.edu/hopper/text?doc=Perseus%3Atext%3A1999.01.0126%3Abook%3D4%-3Achapter%3D8

254   Hdt. 4.36.2 from Herodotus, with an English translation by A. D. Godley. Cambridge. Harvard University Press. 1920 at http://www.perseus.tufts.edu/hopper/text?doc=Perseus%3Atext%3A1999.01.0126%3Abook%3D4%-3Achapter%3D36%3Asection%3D2

255   See https://en.wikipedia.org/wiki/Eratosthenes

256   See https://en.wikipedia.org/wiki/File:Mappa_di_Eratostene.jpg

257   He wrote *"But now that we have examined with sufficient care Ethiopia and the Trogodyte country and the territory adjoining them, as far as the region which is uninhabited because of the excessive heat, and, beside these, the coast of the Red Sea and the Atlantic deep which stretches towards the south, ..."*

258   https://en.wikipedia.org/wiki/Colaeus

259   https://en.wikipedia.org/wiki/Pytheas

260   https://en.wikipedia.org/wiki/Colaeus

261   Hdt. 4.152 Herodotus, with an English translation by A. D. Godley. Cambridge. Harvard University Press. 1920. at http://www.perseus.tufts.edu/hopper/text?doc=Perseus%3Atext%3A1999.01.0126%3Abook%3D4%-3Achapter%3D152

262   https://en.wikipedia.org/wiki/Tartessos

263   http://platoproject.gr/project/the-importance-of-accurate-translation/

264   https://el.wikipedia.org/wiki/%CE%A0%CE%AD%CE%BB%CE%B1%-CE% B3%CE%BF%CF%82

265   Plat. Phaedo 112/3 from Plato in Twelve Volumes, Vol. 1 translated by Harold North Fowler; Introduction by W.R.M. Lamb. Cambridge, MA, Harvard University Press; London, William Heinemann Ltd. 1966 at http://www.perseus.tufts.edu/hopper/text?doc=Perseus%3Atext%3A1999.01.0170%-3Atext%3DPhaedo%3Apage%3D112

266   Thalassa was originally was the primeval spirit of the sea but was later used as another word meaning sea, generally the sea in contrast to the land, but it can also be used for a particular sea or lake, e.g. the sea of Galilee (Tiberias), the Red Sea. See http://biblehub.com/greek/2281.htm

267   Πᾶν πέλαγος

268   Critias 109a

269   Online Etymology Dictionary at http://etymonline.com/index.php?term=pan-. See also https://en.wikipedia.org/wiki/Pan

270   Hdt. 2.23, from The History Of Herodotus by Herodotus, translated by Macaulay, G.C. Volume 1 (of 2) at http://www.gutenberg.org/files/2707/2707-h/2707-h.htm#link22H_4_0001

271   Hdt. 4.8, from The History Of Herodotus by Herodotus, translated by Macaulay, G.C. Volume 1 (of 2) at http://www.gutenberg.org/files/2707/2707-h/2707-h.htm#link22H_4_0001

272   Meteorology translated by E. W. Webster

273   htp://classics.tmit.edu/Aristotle/meteorology.2.ii.html

274   Aristotle, also, exhibits a knowledge of the fact that there is a large sea beyond the pillars of Heracles. See Meteorology by Aristotle translated by E. W. Webster at http://classics.mit.edu/Aristotle/meteorology.2.ii.html

275   See also https://en.wikipedia.org/wiki/Pytheas#cite_note-10

276   Heracles, of the pillars of Heracles, is the Greek form of the Phoenician god Melkart.

277   See https://en.wikipedia.org/wiki/Cassiterides

278   See https://en.wikipedia.org/wiki/Carthage and https://en.wikipedia.org/wiki/Sicilian_Wars

279   Strab. 3.5.11 from The Geography of Strabo. Literally translated, with notes, in three volumes. London. George Bell & Sons. 1903. at http://www.perseus.tufts.edu/hopper/text?doc=Perseus%3Atext%3A1999.01.0239%3Abook%-3D3%3Achapter%3D5%3Asection%3D11 . See also http://penelope.uchicago.edu/Thayer/e/roman/texts/strabo/3e*.html

280   https://en.wikipedia.org/wiki/Spartel

281   It has been proposed that Spartel Bank is, in fact, the site of Atlantis.

282   Gracia, F. J., et al. "Diapiric uplift of an MIS 3 marine deposit in SW Spain: Implications for Late Pleistocene sea level reconstruction and palaeogeography of the Strait of Gibraltar." *Quaternary Science Reviews* 27.23 (2008): 2219-2231.

283   In his article Gutscher, M-A. "The great Lisbon earthquake and tsunami of 1755: lessons from the recent Sumatra earthquakes and possible link to Plato's Atlantis." *European Review* 14.2 (2006): 181-191.

284   Silva, P. G., et al. "Seismic palaeogeography of coastal zones in the Iberian Peninsula: Understanding ancient and historic earthquakes in Spain." *Cuaternario y Geomorfología* 29.1-2 (2015): 31-56.

285   See http://atlantipedia.ie/samples/continent/ (Accessed on 09 April 2018)

286   See also https://en.wikipedia.org/wiki/Continent#The_word_continent (Accessed on 16 April 2018)

287   See https://en.wiktionary.org/wiki/continent (Accessed on 09 April 2018)

288   https://en.wikipedia.org/wiki/Continent (Accessed on 16 April 2018)

289   https://en.wikipedia.org/wiki/Continent#The_word_continent (Accessed on 16 April 2018)

290   The relevant passage reads "... *now that the whole of the other continent was subject to him,* [he] *had crossed over to their continent by a bridge thrown across the neck of the Bosporus ...*"

291   Herodotus, with an English translation by A. D. Godley. Cambridge. Harvard University Press. 1920. At http://www.perseus.tufts.edu/hopper/text?doc=Perseus%3Atext%3A1999.01.0125%3Abook%3D4%3Achapter%-3D118%3Asection%3D1

292   http://www.perseus.tufts.edu/hopper/morph?l=h%29%-2Fpeiron&la=greek&can=h%29%2Fpeiron0&prior=th\n#lexicon nd http://www.perseus.tufts.edu/hopper/morph?l=h%29%-2Fpeiron&la=greek&can=h%29%2Fpeiron0&prior=th\n#lexicon

293   At http://classics.mit.edu/Herodotus/history.4.iv.html

294   See also Hdt. 4.45, Herodotus, with an English translation by A. D. Godley. Cambridge. Harvard University Press.1920. at http://www.perseus.tufts.edu/hopper/text?doc=Hdt.+4.45&fromdoc=Perseus%3Atext%-3A1999.01.0126 and http://www.gutenberg.org/files/2707/2707-h/2707-h.htm#link42H_4_0001

295   https://en.wikipedia.org/wiki/Continent#The_word_continent

296   See http://atlantipedia.ie/samples/continent/ (Accessed on 09 April 2018)

297   *On the Eternity of the World* by Philo, tr. by Yonge (1854/55) at http://www.earlyjewishwritings.com/text/philo/book35.html

298   *On the Eternity of the World* by Philo, in Colson, F. H. "Philo in Ten Volumes (and Two Supplementary Volumes). Vol. IX. Loeb Classical Library." (1985).

299   http://www.yourguidetoitaly.com/origin-of-the-name-italy.html (Accessed on 22 Nove,ber 2018)

300   From Plato in Twelve Volumes, Vol. 9 translated by W.R.M. Lamb. Cambridge, MA, Harvard University Press; London, William Heinemann Ltd. 1925. at http://www.perseus.tufts.edu/hopper/text?doc=Perseus%3Atext%-3A1999.01.0180%3Atext%3DTim.%3Asection%3D25a

301   See http://www.gutenberg.org/files/1572/1572-h/1572-h.htm

302   See http://www.perseus.tufts.edu/hopper/text?doc=Perseus%3Atext%-3A1999.01.0180%3Atext%3DTim.%3Asection%3D25a (Accessed on 09 April 2018)

303   See http://www.perseus.tufts.edu/hopper/morph?l=h%29pei%-2Frou&la=greek&can=h%29pei%2Frou0&prior=th=s#Perseus:text:1999.04.0057:entry=h)peiro/w-contents

304   Hdt. 4.43, The History Of Herodotus by Herodotus, translated by Macaulay, G.C. Volume 1 (of 2) at http://www.gutenberg.org/files/2707/2707-h/2707-h.htm#link42H_4_0001 (Accessed on 09 April 2018)

305   The Greek word used for 'continent' (Aelian wrote in Greek) was also ἤπειρον / ípeiron. From *Claudii Aeliani de natura animalium libri xvii, varia historia, epistolae, fragmenta,* Vol 2. Aelian. Rudolf Hercher. In Aedibus B.G. Teubneri. Lipsiae. 1866 at http://www.perseus.tufts.edu/hopper/text?doc=Perseus%-3Atext%3A2008.01.0591%3Abook%3D3%3Achapter%3D18 (Accessed on 22 November 2018)

306   *Claudius Ælianus his various history.* For Thomas Dring translated by T Stanley (1665) Book III, Chapter 18

307   See http://atlantipedia.ie/samples/continent/ (Accessed on 09 April 2018) He refers to "*[Timaeus 24e, 25a, 25d Critias 108e, 113c, 113d, 113e, 114a, 114b, 114e, 115b, 115e, 116a, 117c, 118b, 119c]*"

308   See http://atlantipedia.ie/samples/size-of-atlantis/ (Accessed on 09 April 2018)

309   http://www.perseus.tufts.edu/hopper/morph?l=mei%-2Fzwn&la=greek&can=mei%2Fzwn0&prior=*)asi/as&d=Perseus:text:1999.01.0179:text=Tim.:section=24e&i=1#lexicon

310   LSJ lexicon

311   Middle Liddell lexicon

312   http://atlantipedia.ie/samples/asia/

313   http://atlantipedia.ie/samples/libya/

314   See Hdt. 4.42.1, Herodotus, with an English translation by A. D. Godley. Cambridge. Harvard University Press. 1920 at http://www.perseus.tufts.edu/hopper/text?doc=Perseus%3Atext%3A1999.01.0126%3Abook%3D4%-3Achapter%3D42%3Asection%3D1 and *The History of Herodotus* translated by G Rawlinson at http://classics.mit.edu/Herodotus/history.4.iv.html .

315   https://en.wikipedia.org/wiki/Rioni_River (Accessed on 5 March 2018)

316   https://en.wikipedia.org/wiki/Kerch_Strait (Accessed on 5 March 2018)

317   See https://goo.gl/vRzLYx (Accessed 3 March 2018)

318   See https://goo.gl/ShhN8w (Accessed on 5 March 2018)

319   See https://goo.gl/USPFf7 (Accessed on 5 March 2018)

320   See https://goo.gl/3YcPe1 (Accessed on 5 March 2018)

321   Vinci, Felice. *The Baltic Origins of Homer's Epic Tales: The Iliad, the Odyssey, and the Migration of Myth.* Simon and Schuster, 2005.

322   Who lived after Plato and in the first century BCE. See https://en.wikipedia.org/wiki/Diodorus_Siculus (Accessed on 5 March 2018)

323   Diodorus Siculus, when describing the size of Great Britain, states that "*Of the sides of Britain the shortest, which extends along Europe, is seven thousand five hundred stades, the second, from the Strait to the (northern) tip, is fifteen thousand stades, and the last is twenty thousand stades, so that the entire circuit of the island amounts to forty-two thousand five hundred stades.*" Taking the Itinerary stade (or stadion), which was used for journeys and was about 157 metres long, this means that the circumference, or 'length', of Britain is, according to Diodorus Siculus, about 6 670 kilometres.

324   Shuckburgh, Evelyn S., ed. *The Histories of Polybius: Translated from the Text of F. Hultsch.* Vol. 2. from http://www.gutenberg.org/ebooks/44126 (Accessed on 7 September 2018)

325   http://ports.com

326   As calculated on Herodotus' definitions

327   Whilst, nevertheless, acknowledging that there are many uncertainties regarding the starting and end points of the routes taken [and even of the actual routes] in calculating the length of Europe/Libya and Asia and even the distance around Hatton Rockall.

328   See perseus.tufts.edu - http://goo.gl/vUX3my (Accessed on 13 March 2015)

329   See http://logeion.uchicago.edu/index.html#%CF%80%CF%81%CF%8C (Accessed on 13 March 2015)

330   The Online Etymology Dictionary [http://www.etymonline.com/index.php?term=pro- (Accessed on 13 March 2015)] discusses various cognates of "*pro*" and links it (both the Greek and Latin versions) to "*pro*" in the Pre-Indo-European language and also to the root "*per*". *Per* signifies, amongst other meanings, "forward" or "through". [See the etymology given of 'per' at http://www.etymonline.com/index.php?term=per&allowed_in_frame=0]

331   Kershaw, Stephen P. *A Brief History of Atlantis: Plato's Ideal State.* Robinson, 2017.

332   Personal communication

333   See http://en.wikipedia.org/wiki/Temperate_climate (Accessed on 30 October 2014)

334   *Perseus Critias*, par. 118b

335   Based on map obtained from http://maps.ngdc.noaa.gov/viewers/bathymetry/ (Accessed on 24 October 2014) The original map is United States Government material and is not subject to copyright protection within the United States and is in the public domain.

336   See https://en.wikipedia.org/wiki/Norse_colonization_of_North_America (Accessed on 10 September 2018)

337   I shall deal with their climates when I deal more fully with the issue of climate in Section 4.

338   Apart from areas of continental shelf immediately surrounding the continents.

339   Meyer, Romain, J. van Wijk, and L. Gernigon. "The North Atlantic Igneous Province: A review of models for its formation." *Geological Society of America Special Papers* 430 (2007): 525-552.

340   Morton, A. C., et al. "Detrital zircon age constraints on the provenance of sandstones on Hatton Bank and Edoras Bank, NE Atlantic." *Journal of the Geological Society* 166.1 (2009): 137-146.

341   Smith, L. K., R. S White, N. J. Kusznir, & iSIMM Team "Structure of the Hatton Basin and adjacent continental margin." (2005) in: Dore, A. G. & B. A. Vining, (eds) Petroleum Geology: North-West Europe and Global Perspectives—Proceedings of the 6th Petroleum Geology Conference, 947–956.

342   See also Neish, J. K.. "Seismic structure of the Hatton–Rockall area: an integrated seismic/modelling study from composite datasets." *Geological Society, London, Petroleum Geology Conference series.* Vol. 4. Geological Society of London, 1993., Roberts, D. G. "Marine geology of the Rockall Plateau and Trough." *Philosophical Transactions of the Royal Society of London. Series A, Mathematical and Physical Sciences* 278.1285 (1975): 447-509., *doi: 10.1098/rsta.1975.0033,* White, R. S., and L. K. Smith. "Crustal structure of the Hatton and the conjugate east Greenland rifted volcanic continental margins, NE Atlantic." *Journal of Geophysical Research: Solid Earth (1978–2012)* 114.B2 (2009).

343   Boldreel, L. O. and M. S. Andersen. "Tertiary development of the Faeroe-Rockall Plateau based on reflection seismic data." *Bulletin of the Geological Society of Denmark* 41.2 (1994): 162-180.

344   See also Roberts, D G Appendix II Site Survey in Hatton-Rockall Basin (Sites 116 and 117),

345   From http://en.wikipedia.org/wiki/File:World_geologic_provinces.jpg (Accessed on 26 September 2014) The original map was produced by the U.S. Geological Survey (http://www.usgs.gov), is United States Government material and is not subject to copyright protection within the United States and is in the public domain.

346   See section 4.2.1.2 in this regard.

347   Backman, J.. "Miocene-Pliocene nannofossils and sedimentation rates in the Hatton-Rockall basin, NE Atlantic Ocean." (1980). (an inaugural dissertation) at pp 75-77 and Laughton, A. S., W.A. Berggren, et al., - Initial Reports of the Deep Sea Drilling Project, 12 (U.S. Government Printing Office) (1972) at p 422

348   See also Pearson, I., and D. G. Jenkins. "Unconformities in the Cenozoic of the North-East Atlantic." *Geological Society, London, Special Publications* 21.1 (1986): 79-86.

349   i.e. since about 10 000 BCE

350   Roberts, D. G. "Marine geology of the Rockall Plateau and Trough." *Philosophical Transactions of the Royal Society of London. Series A, Mathematical and Physical Sciences* 278.1285 (1975): 447-509 and Wickham-Jones, C. R., and S. Dawson. "The scope of Strategic Environmental Assessment of North

Sea Area SEA7 with regard to prehistoric and early historic archaeological remains." *www. offshore-sea. org. uk/site/index. Php* (2006).

351 Watts, A. B., B. C. Schreiber, and D. Habib. "Dredged rocks from Hatton Bank, Rockall Plateau." *Journal of the Geological Society* 131.6 (1975): 639-646.

352 Summary and Conclusions, p. 1188 (doi:10.2973/dsdp.proc.12.121.1972) of the Deep Sea Drilling Project accessed at http://www.deepseadrilling.org/12/dsdp_toc.htm on 19 September 2014.

353 See http://en.wikipedia.org/wiki/Stadion_%28unit_of_length%29 (Accessed on 21 October 2014)

354 Based on bathymetry map obtained from http://maps.ngdc.noaa.gov/viewers/bathymetry/ (Accessed on 21 October 2014) The original map is United States Government material and is not subject to copyright protection within the United States and is in the public domain.

355 Roberts, D. G. "Marine geology of the Rockall Plateau and Trough." *Philosophical Transactions of the Royal Society of London. Series A, Mathematical and Physical Sciences* 278.1285 (1975): 447-509. See also http://en.wikipedia.org/wiki/Nautical_miles (Accessed on 19 October 2014) for conversion factors from nautical miles.

356 Which has a surface area of 70 273 km2 See http://en.wikipedia.org/wiki/Republic_of_Ireland (Accessed on 19 October 2014)

357 Which has a surface area of 103 001 km2 See http://en.wikipedia.org/wiki/Iceland (Accessed on 19 October 2014)

358 See http://en.wikipedia.org/wiki/Geography_of_the_United_Kingdom (Accessed on 19 October 2014)

359 Based on bathymetry map obtained from http://maps.ngdc.noaa.gov/viewers/bathymetry/ (Accessed on 21 October 2014) The original map is United States Government material and is not subject to copyright protection within the United States and is in the public domain.

360 See http://en.wikipedia.org/wiki/List_of_mountains_and_hills_of_the_United_Kingdom (Accessed on 24 October 2014)

361 Or 4,409 ft

362 See http://en.wikipedia.org/wiki/Carrauntoohil (Accessed on 24 October 2014)

363 Or 3,406 ft

364 Obtained from http://maps.ngdc.noaa.gov/viewers/bathymetry/ (Accessed on 19 October 2014) The original map is United States Government material and is not subject to copyright protection within the United States and is in the public domain.

365 At http://www.ngdc.noaa.gov/mgg/topo/globega2.html and http://www.ngdc.noaa.gov/mgg/topo/pictures/GLOBALsealevelsm.jpg (Accessed on 24 October 2014.) The original map is United States Government material and is not subject to copyright protection within the United States and is in the public domain.

366 And based upon the model used.

367   Although the land areas of the map are in tones of greens and browns it must be borne in mind that, at the time, large areas were, in fact, under ice.

368   See *Timaeus*

369   Based on bathymetry map obtained from <u>http://www.ngdc.noaa.gov/mgg/topo/globega2.html</u> (Accessed on 24 October 2014) The original map is United States Government material and is not subject to copyright protection within the United States and is in the public domain.

370   Based on bathymetry map obtained from <u>http://www.ngdc.noaa.gov/mgg/topo/globega2.html</u> (Accessed on 24 October 2014) The original map is United States Government material and is not subject to copyright protection within the United States and is in the public domain.

371   *Critias*

372   See also the *Perseus Critias*, at par. 118C - D

373   *Critias*

374   See also *Perseus Critias*, at par. 118D

375   *Critias*

376   See also *Perseus Critias*, at par. 119a

377   Whether the reference to the "entire country" encompassed only the plain itself or whether it extended beyond the plain one does not know but, based upon the measurements given by Plato of the plain (3 000 by 2 000 stadia), the size of the allotments (10 by 10 stadia) and simple arithmetic, it appears that the number of "lots" is consistent with the size of the plain.

378   Roberts, D. G. "Marine geology of the Rockall Plateau and Trough." *Philosophical Transactions of the Royal Society of London. Series A, Mathematical and Physical Sciences* 278.1285 (1975): 447-509 at p 458

379   As I stated above the measurements of the plain were, depending upon the conversion factor used (and obviously depending from where it was measured), between 471 km by 314 km or 627 km by 418 km.

380   It is possible that the plain distorted and shrunk as Hatton Rockall sunk but, if this is or may be so, it is beyond my abilities to calculate to what extent (if at all) it did so and, accordingly, I leave this for others to consider.

381   Roberts, D. G. "Marine geology of the Rockall Plateau and Trough." *Philosophical Transactions of the Royal Society of London. Series A, Mathematical and Physical Sciences* 278.1285 (1975): 447-509

382   Roberts, D. G. Marine Geology of the Rockall Plateau and Trough *supra*

383   Davies, T. A. and A. S. Laughton. "Sedimentary processes in the North Atlantic." *Initial reports of the deep sea drilling project* 12 (1972): 905-934. (Accessed at <u>http://www.deepseadrilling.org/12/dsdp_toc.htm</u> on 19 September 2014.)

384   Sayago-Gil, M., D. Long, K. Hitchen, V. Díaz-del-Río, L.M. Fernández-Salas, P. Durán-Muñoz in Geomorphology of the western slope of Hatton Bank (Rockall Plateau, NE Atlantic Ocean) revealed by multibeam bathymetry and high-resolution seismic data: control by bottom current regime (2009) from *nora.nerc.ac.uk/9492/1/Sayago-Gil_et_al_v1.pdf* - NERC Open Access Research Archive (NORA) Available: <u>http://nora.nerc.ac.uk/</u> (Accessed on 20 October 2014)

385   Obtained from http://www.seismicatlas.org/entity?id=bc82829e-e629-
      4036-a7d8-aa7015ded422 - BGSRockall 2001/01_01 VSA Author: Estelle
      Mortimer Created: 2008-02-05. (Accessed on 24 October 2014)

386   Diagram based on map in Davies, T. A., and A. S. Laughton. "Sedimentary
      processes in the North Atlantic." *Initial reports of the deep sea drilling project*"
      12 (1972): 905-934. at p. 923

387   Up to 1300 feet / 400 metres

388   Berndt, C., et al. "Kilometre-scale polygonal seabed depressions in the Hatton
      Basin, NE Atlantic Ocean: Constraints on the origin of polygonal faulting."
      Marine Geology 332 (2012): 126-133. and Jacobs, C. L. "An appraisal of the
      surface geology and sedimentary processes within SEA7, the UK continental
      shelf." (2006). at p 48

389   Berndt, C., et al. "Kilometre-scale polygonal seabed depressions in the Hatton
      Basin, NE Atlantic Ocean: Constraints on the origin of polygonal faulting."
      Marine Geology 332 (2012): 126-133.

390   Rebecca Ross, a marine ecologist at Plymouth University

391   http://thinkingdeepblue.blogspot.com/ (Accessed 17 October 2014)

392   See also the JC060 cruise blog at http://www.eu-hermione.net/rockall-
      hatton-basin (Accessed 17 October 2014) whose author comments that these
      polygons are unique because of their presence at sea floor level.

393   See also Huvenne, V. AI. "RRS James Cook Cruise 60, 09 May-12 Jun 2011.
      Benthic habitats and the impact of human activities in Rockall Trough, on
      Rockall Bank and in Hatton Basin." (2011) and see also Hansen, D. M. and
      Cartwright, J. "The three-dimensional geometry and growth of forced folds
      above saucer-shaped igneous sills." *Journal of Structural Geology* 28.8 (2006):
      1520-1535.

394   The polygons in the photographs are much smaller than the polygons formed
      on Hatton Rockall and are also much smaller than the dimension of the lots as
      given by Plato.

395   Approximately 5 000 – 4 000 BCE. See http://en.wikipedia.org/wiki/List_of_
      prehistoric_lakes

396   At its maximum extent at about 4 200 BCE.

397   or the Mammal Igneous (or Volcanic) Centre.

398   See also comment at http://see-atlas.leeds.ac.uk:8080/entity?id=bc82829e-
      e629-4036-a7d8-aa7015ded422 (Accessed 29 March 2016)

399   Obtained from http://maps.ngdc.noaa.gov/viewers/bathymetry/ (Accessed on
      19 October 2014) The original map is United States Government material and
      is not subject to copyright protection within the United States and is in the
      public domain.

400   Hitchen, K. "The geology of the UK Hatton-Rockall margin. *Marine and
      Petroleum Geology* 21.8 (2004): 993-1012.

401   Hitchen, K. "The geology of the UK Hatton-Rockall margin." *Marine and
      Petroleum Geology* 21.8 (2004): 993-1012.

402   Obtained from http://www.seismicatlas.org/entity?id=6a73b36f-32cd-4014-
      ac02-20ebf16f63bf - Detail of Mammal igneous complex VSA Author: Estelle
      Mortimer Date Created: 2008-02-05 (Accessed on 20 October 2014)

403   Dorschel, B., et al. "Growth and erosion of a cold-water coral covered carbonate mound in the Northeast Atlantic during the Late Pleistocene and Holocene." *Earth and Planetary Science Letters* 233.1 (2005): 33-44.

404   Mienis, F. et al. "Sediment accumulation on a cold-water carbonate mound at the Southwest Rockall Trough margin." *Marine Geology* 265.1 (2009): 40-50.

405   Kano, A. et al. "Age constraints on the origin and growth history of a deep-water coral mound in the northeast Atlantic drilled during Integrated Ocean Drilling Program Expedition 307." *Geology* 35.11 (2007): 1051-1054.

406   Mienis, F. et al. "Sediment accumulation on a cold-water carbonate mound at the Southwest Rockall Trough margin." *Marine Geology* 265.1 (2009): 40-50.

407   Of up to hundreds of thousands of years.

408   Mienis, F. et al. "Sediment accumulation on a cold-water carbonate mound at the Southwest Rockall Trough margin." *Marine Geology* 265.1 (2009): 40-50.

409   Frank, N. et al. "Deep-water corals of the northeastern Atlantic margin: carbonate mound evolution and upper intermediate water ventilation during the Holocene." *Cold-water corals and ecosystems*. Springer Berlin Heidelberg, 2005. 113-133.

410   van der Land, C. et al. "Carbonate mound development in contrasting settings on the Irish margin." *Deep Sea Research Part II: Topical Studies in Oceanography* 99 (2014): 297-306.

411   Mienis, F., et al. "Sediment accumulation on a cold-water carbonate mound at the Southwest Rockall Trough margin." *Marine Geology* 265.1 (2009): 40-50.

412   Frank, N. et al. "Deep-water corals of the northeastern Atlantic margin: carbonate mound evolution and upper intermediate water ventilation during the Holocene." *Cold-water corals and ecosystems*. Springer Berlin Heidelberg, 2005. 113-133.

413   Eisele, M., et al. "Productivity controlled cold-water coral growth periods during the last glacial off Mauritania." *Marine Geology* 280.1 (2011): 143-149.

414   Wienberg, C. et al. "Glacial cold-water coral growth in the Gulf of Cádiz: Implications of increased palaeo-productivity." *Earth and Planetary Science Letters* 298.3 (2010): 405-416.

415   Belka, Z. "Early Devonian Kess-Kess carbonate mud mounds of the eastern Anti-Atlas (Morocco), and their relation to submarine hydrothermal venting." *Journal of Sedimentary Research* 68.3 (1998).

416   See also section 6.3 where reference is made to the possibility that the Atlas and Anti-Atlas mountains are supported by underlying magma.

417   Pirlet, H. et al. "Diagenetic formation of gypsum and dolomite in a cold water coral mound in the Porcupine Seabight, off Ireland." *Sedimentology* 57.3 (2010): 786-805.

418   Stoker, M. S., et al. "Eocene post-rift tectonostratigraphy of the Rockall Plateau, Atlantic margin of NW Britain: linking early spreading tectonics and passive margin response." *Marine and Petroleum Geology* 30.1 (2012): 98-125.

419   See http://www.bbc.com/news/uk-scotland-29539119 (Accessed on 12 August 2015)

420   If it was not there would be no debate.

421   Whilst I suggest that Rockall Islet might have sunk, it is beyond my expertise and knowledge to state what proportion, if any, is due to better equipment and what is due to actual sinking. I do, however, suggest that it is also improbable that the measurements taken in 1977 were so inaccurate as to allow for such a large discrepancy.

422   https://en.wikipedia.org/wiki/Climate (Accessed on 23 June 2015)

423   http://www.ipcc.ch/publications_and_data/ar4/wg1/en/faq-1-1.html (Accessed on 23 June 2015)

424   https://en.wikipedia.org/wiki/LOWERN (Accessed on 23 June 2015)

425   See also http://www.climateandweather.net/global-warming/factors-that-influence-climate.html (Accessed on 23 June 2015)

426   http://www.ncdc.noaa.gov/paleo/milankovitch.html (Accessed on 23 June 2015)

427   http://www.ipcc.ch/publications_and_data/ar4/wg1/en/ch1s1-4-2.html (Accessed on 23 June 2015)

428   See http://www.ondrejdanek.cz/rockall/en/rockall.html and http://www.noc.soton.ac.uk/obe/PROJECTS/EEL/weather.php See also fn. 20

429   See http://www.ondrejdanek.cz/rockall/en/rockall.html and Page 2.27 of 1919R004Annexs2Physicalandchemicalaspects_f[1].pdf dated October 2008 obtained at dcen.gov.ie - http://www.dcenr.gov.ie/NR/rdonlyres/2A154470-B6B1-495B-B268-D5429B747F13/0/1673R002Annexs2Physicalandchemical aspects_final.pdf (Accessed on 16 June 2011)

430   See page 2.25 of 1919R004Annexs2Physicalandchemicalaspects_f[1].pdf referred to in the previous footnote See also http://oceancurrents.rsmas.miami.edu/atlantic/north-atlantic-drift.html

431   From http://en.wikipedia.org/wiki/File:Golfstream.jpg by RedAndr (Accessed on 24 October 2014) The file is licensed under the Creative Commons Attribution-Share Alike 4.0

432   Based on map from: http://www.noc.soton.ac.uk/obe/PROJECTS/EEL/whyrockall.php (Accessed on 24 October 2014)

433   Li, C. and D. S. Battisti. "Reduced Atlantic storminess during Last Glacial Maximum: Evidence from a coupled climate model." *Journal of Climate* 21.14 (2008): 3561-3579.

434   Alonso-Garcia, M., F. J. Sierro, and J. A. Flores. "Arctic front shifts in the subpolar North Atlantic during the Mid-Pleistocene (800–400ka) and their implications for ocean circulation." *Palaeogeography, Palaeoclimatology, Palaeoecology* 311.3 (2011): 268-280.

435   Zhirov, N. *Atlantis: Atlantology: Basic Problems*. University Press of the Pacific, 2001. at pp 253 and 336

436   Ezat, M. M., T. L. Rasmussen, and J. Groeneveld. "Persistent intermediate water warming during cold stadials in the southeastern Nordic seas during the past 65 ky." *Geology* 42.8 (2014): 663-666.

437   Thornalley, D. J. R., et al. "A warm and poorly ventilated deep Arctic Mediterranean during the last glacial period." *Science* 349.6249 (2015): 706-710.

438   See http://en.wikipedia.org/wiki/Last_Glacial_Maximum

439   See http://en.wikipedia.org/wiki/Late_Glacial_Maximum

440   Diefendorf, A. F. *Late-Glacial to Holocene climate variability in western Ireland.* Diss. University of Saskatchewan, 2005.

441   See also O'Connell, M., C. C. Huang, and U. Eicher. "Multidisciplinary investigations, including stable-isotope studies, of thick Late-glacial sediments from Tory Hill, Co. Limerick, western Ireland." *Palaeogeography, Palaeoclimatology, Palaeoecology* 147.3 (1999): 169-208.

442   Before the present – often measured from 1950.

443   i.e. between approximately 10 800 BCE and 9 500 BCE

444   See http://en.wikipedia.org/wiki/Younger_Dryas

445   Wolbach, Wendy S., et al. "Extraordinary biomass-burning episode and impact winter triggered by the Younger Dryas cosmic impact⊠ 12,800 years ago. 2. Lake, marine, and terrestrial sediments." *The Journal of Geology* 126.2 (2018) and Wolbach, Wendy S., et al. "Extraordinary Biomass-Burning Episode and Impact Winter Triggered by the Younger Dryas Cosmic Impact⊠ 12,800 Years Ago. 1. Ice Cores and Glaciers."

446   Rasmussen, T. L., E. Thomsen, and M. Moros. "North Atlantic warming during Dansgaard-Oeschger events synchronous with Antarctic warming and out-of-phase with Greenland climate." *Scientific Reports* **6**, 20535 (2016)

447   See also CAGE - Center for Arctic Gas Hydrate, Climate and Environment. "'Ice age blob' of warm ocean water discovered south of Greenland." ScienceDaily. ScienceDaily, 19 February 2016. <www.sciencedaily.com/releases/2016/02/160219134816.htm>. (Accessed on 22 February 2016)

448   Hu, Aixue, et al. "Influence of Bering Strait flow and North Atlantic circulation on glacial sea-level changes." *Nature Geoscience* 3.2 (2010): 118-121.

449   Elias, S.A. et al., 1996. "Life and times of the Bering land bridge." Nature 382, 60–63.

450   Between 14 000 and 11 000 years BP

451   Rasmussen, S. O., et al. "A new Greenland ice core chronology for the last glacial termination." *Journal of Geophysical Research: Atmospheres* (1984–2012) 111.D6 (2006).

452   Alley, R. B. "The Younger Dryas cold interval as viewed from central Greenland." *Quaternary Science Reviews* 19.1 (2000): 213-226.

453   11 500 BP

454   See however also Rasmussen, S. O., et al. "A new Greenland ice core chronology for the last glacial termination." *Journal of Geophysical Research: Atmospheres* (1984–2012) 111.D6 (2006) who put it at 11 703 b2k.

455   Alley, R.B, "The Younger Dryas cold interval as viewed from central Greenland" *supra*, Alley, R. B. "Ice-core evidence of abrupt climate changes." *Proceedings of the National Academy of Sciences* 97.4 (2000): 1331-1334. and Taylor, K. C., et al. "The Holocene-Younger Dryas transition recorded at Summit, Greenland." *Science* 278.5339 (1997): 825-827. According to the latter article a change of in the order of 7.5 °C could have occurred in less than 15 years.

456   See also Li, C., et al. "Abrupt climate shifts in Greenland due to displacements of the sea ice edge." *Geophysical Research Letters* 32.19 (2005)., Dansgaard–Oeschger event. See http://en.wikipedia.org/wiki/Dansgaard-Oeschger_event#cite_note-2 and Rahmstorf, S. "Timing of abrupt climate change: A precise clock." *Geophysical Research Letters* 30.10 (2003).

457   http://en.wikipedia.org/wiki/Ireland#Geography

458   See Wikipedia for a discussion of oceanic climates http://en.wikipedia.org/wiki/Oceanic_climate

459   http://en.wikipedia.org/wiki/Republic_of_Ireland

460   http://en.wikipedia.org/wiki/Ireland

461   *Arbutus unedo* is also known as the Irish or Killarney Strawberry tree

462   http://en.wikipedia.org/wiki/County_Kerry

463   http://en.wikipedia.org/wiki/Ireland (Accessed 6 June 2012)

464   The History of Ireland by Geoffrey Keating (Foras Feasa ar Éireann le Seathrún Céitinn) Translated into English by Edward Comyn and Patrick S. Dinneen

465   http://en.wikipedia.org/wiki/Climate_of_south-west_England

466   http://en.wikipedia.org/wiki/Geography_of_Cornwall#Climate

467   http://en.wikipedia.org/wiki/Geography_of_Cornwall#Climate

468   http://en.wikipedia.org/wiki/Devon

469   http://en.wikipedia.org/wiki/Wales

470   http://en.wikipedia.org/wiki/Faroe_Islands

471   *Critias* See also *Perseus Critias* at par's 113e to 114b

472   In the Perseus *Critias* translation, at par. 114, this is translated as " … *his younger twin-brother, who had for his portion the extremity of the island near the pillars of Heracles up to the part of the country now called Gadeira …*".

473   It appears that 'Gades' is the Latin derived and 'Gadeira' is the Greek derived form of the word. See http://en.wikipedia.org/wiki/Cadiz . (Accessed on 20 September 2014)

474   It is not always the specific words 'Gades' or 'Gadeira' which are accepted as being a reference to Cadiz. In the translation of Josephus' Antiquities of the Jews [Josephus, Flavius. "Josephus Complete Works. trans. by William Whiston." *Grand Rapids, MI: Kregel* (1960- Reprinted 1976)] reference is made (in the English translation), at Verse 1, Chapter VI, Book 1 and in the context of events after the Noachian flood, to Cadiz. This, Cadiz, is the translation given of the Greek word "Γαδείρων" (or Gadeiron) and of the Latin word "gazitorum". This reference is made in the context of a reference to the western most (Cadiz) and eastern most (Tanais) extremes in which people had allegedly settled after the flood. Another translation of Josephus' work may be found at https://sites.google.com/site/latinjosephus/antiquities/book-1 (Accessed on 3 April 2018)

475   The original language of the inhabitants of Atlantis, ancient Egyptian, ancient Greek and English.

476   I was able, in a short search on the internet, to find four different cities or towns called Cadiz around the world.

477   In Dr. Smith's "*Dictionary of Ancient Geography*.", written in the early 1800's, he states that " *'Fortunatæ Insulæ' is one of those geographical names whose origin*

is lost in mythic darkness, but which afterwards came to have a specific application, so closely resembling the old mythical notion, as to make it almost impossible to doubt that that notion was based, in part at least, on some vague knowledge of the regions afterwards discovered. The earliest Greek poetry places the abode of the happy departed spirits far beyond the entrance of the Mediterranean, at the extremity of the earth, and upon the shores of the river Oceanus, or in islands in its midst; ..."

478   Adapted from map obtained from http://maps.ngdc.noaa.gov/viewers/bathymetry/ (Accessed on 19 October 2014) The original map is United States Government material and is not subject to copyright protection within the United States and is in the public domain.

479   See http://en.wikipedia.org/wiki/Herodotus

480   The *History of Herodotus by Herodotus*, Book 1, Clio Translated by George Rawlinson. See also, for an alternate translation, *The History of Herodotus*, by Herodotus. Translated into English by G. C. Macaulay.

481   It has been suggested that, according to Irish mythology, the Irish are possibly descended from Scythian colonists.

482   That Gadeira/Gades was an island becomes unambiguously apparent from the work of Appollodorus, the '*Library of Appollodorus*', referred to below.

483   See Wikipedia http://en.wikipedia.org/wiki/Eratosthenes

484   He does, however, criticise certain aspects of it.

485   This he says is not correct because another author, Artemidorus, states that they are only 1700 stadia apart. It is clear from this that, even at the stage at which Strabo wrote (which was about 400 years after Plato), Gades was associated with the modern day Cadiz.

486   *Geographica* or *The Geography of Strabo* translated by H. L. Jones, in the Loeb Classical Library edition Vol. 2 of 8: Harvard University Press 1923 3.2.11

487   See http://en.wikipedia.org/wiki/Sacred_promontory (Accessed on 27 December 2014)

488   The other two are Holyhead, Wales and Hook Head, Wexford, Ireland.

489   McGrail, Seán. *Ancient Boats in North-West Europe: The archaeology of water transport to AD 1500*. Routledge, 2014. at page 263

490   Although it is generally referred to as the *Library of Appollodorus* the attribution to Appollodorus is now regarded as false.

491   See Wikipedia, http://en.wikipedia.org/wiki/Bibliotheca_(Pseudo-Apollodorus)

492   http://goo.gl/D0QSpl (Accessed on 20 September 2014)

493   Also known as Pherecydes of Leros. See http://en.wikipedia.org/wiki/Pherecydes_of_Athens (Accessed on 20 September 2014).

494   Book 2 of the *Library by Appollodorus*; Appollodorus, and Appollodorus (of Athens.). *The Library [by] Appollodorus*. Ed. Sir James George Frazer. Harvard University Press, 1921 accessible at http://www.theoi.com/Text/Apollodorus2.html (Accessed on 20 September 2014)

495   Also translated as Mount Abas – see http://goo.gl/Zk2o7K (perseus.tufts.edu) (Accessed on 20 September 2014)

496   Paus. 1.35.8 from Pausanias. Pausanias Description of Greece with an English Translation by W.H.S. Jones, Litt.D., and H.A. Ormerod, M.A., in 4 Volumes. Cambridge, MA, Harvard University Press; London, William Heinemann Ltd. 1918 at http://www.perseus.tufts.edu/hopper/text?doc=Perseus%3Atext%3A1999.01.0160%3Abook%3D1%3Achapter%-3D35%3Asection%3D8 (Accessed on 28 May 2018)

497   See http://en.wikipedia.org/wiki/Strabo (Accessed on 25 October 2014)

498   *Geographica* or *The Geography of Strabo* translated by H. L. Jones, in the Loeb Classical Library edition Vol. 2 of 8: Harvard University Press 1923

499   Baetis is the name given by the Phoenicians, it is currently known as the Guadalquiver

500   Homer's *Iliad* viii. 485 [See Project Gutenberg's The Iliad of Homer translated by Alexander Pope, with notes by the Rev. Theodore Alois Buckley 1899] [Buckley, Theodore Alois. *The Odyssey of Homer*. Burt, 1899. - http://www.gutenberg.org/ebooks/6130 (Accessed on 3 April 2018)

501   McGee, Thomas D'Arcy. *A Popular History of Ireland: From the Earliest Period to the Emancipation of the Catholics (Complete)*. (1869) Vol. 1. Library of Alexandria.

502   I hope to provide, through a possible etymology of the word Gades/Gadeira, corroboration for my hypothesis that Plato's reference to Gades is a reference to Ireland.

503   James Hardiman in his work *The history of the town and county of the town of Galway. From the earliest period to the present time.* Dublin, 1820 [Hardiman, James. *The History of the Town and Country of the Town of Galway: From the Earliest Period to the Present Time... W. Folds, 1820.] stated, at Part 1, Chapter 1, page 1, that "*The general opinion concerning etymological enquiries seems to be, that they are rather curious than useful; at the same time it stands confessed, that, in many instances, such disquisitions may become material and interesting, particularly should they lead to the establishment or corroboration of historical facts or tend to illustrate the ancient state of the places under investigation.*"

504   There are various spellings of the name Milesius, most having a Latin structure or sound to them, but he is even referred to by the name of Milidh. See *The History of Ireland by Geoffrey Keating* as translated into English by E. Comyn and P. S. Dinneen (Published by the Ex-classics Project, 2009 http://www.exclassics.com, http://www.exclassics.com/ceitinn/forintro.htm (Accessed on 20 September 2014) and Keating, Geoffrey. "The General History of Ireland, trans." *Dermod O'Connor (Dublin, 1723)* (1841).

505   "Míl Espáine" See Wikipedia Milesians (Irish) http://en.wikipedia.org/wiki/Milesians_(Irish)

506   It may also be noted (inasmuch as it is suggested that there are links between these various waves of invaders and that there is a Greek link or background) that a powerful ancient Greek city was the Ionian city of Miletus. [See https://en.wikipedia.org/wiki/Miletus (Accessed on 3 April 2018)] Its inhabitants are known, in English, as Milesians and it was known for, amongst other things, its maritime prowess and its colonies. [It had about 90 colonies around the

Mediterranean extending as far as present day Spain. See https://en.wikipedia.org/wiki/Colonies_in_antiquity#Greek_colonies (Accessed on 3 April 2018)]

507   Also known as Gaidel Glas

508   See Wikipedia, Goídel Glas, http://en.wikipedia.org/wiki/Goidel_Glas (Accessed on 20 September 2014)

509   Holinshed, Raphael, *The Chronicles of England, Scotland and Ireland*. London : Printed by Henry Denham, at the expenses of Iohn Harison, George Bishop, Rafe Newberie, Henrie Denham, and Thomas Woodcocke, [1587].

510   Keating, Geoffrey. "*The General History of Ireland, trans.*" *Dermod O'Connor (Dublin, 1723)* (1841) at p. 113

511   O'Flaherty, Roderic. "*Ogygia, or, a chronological account of Irish events:: collected from very ancient documents,...* Written originally in Latin by Roderic O'Flaherty, Esq. Translated by the Revd. James Hely, AB..." (1793)

512   p. 73

513   p. 123

514   As an aside there still exists the ancient clan/surname name O'Gadrha which originates from Connacht or Connaught. The "O' " means son, or grandson, of Gadrha. The history of this clan name goes back to prehistory and the exact meaning of Gadrha is lost in the mists of time. Connacht itself has an ancient history and is one of the original Five Provinces of Ireland. What the root of this name is I do not know.

515   Holinshed, Raphael, *The Chronicles of England, Scotland and Ireland*. London : Printed by Henry Denham, at the expenses of Iohn Harison, George Bishop, Rafe Newberie, Henrie Denham, and Thomas Woodcocke, [1587].

516   See also http://en.wikipedia.org/wiki/Holinshed's_Chronicles ,The Holinshed Project: http://english.nsms.ox.ac.uk/holinshed/, http://goo.gl/nHC07S (sceti. library.upenn.edu) and http://en.wikipedia.org/wiki/Raphael_Holinshed

517   "*Portingall*"

518   See Fernández Domínguez, E, et al. "Ancient DNA analysis of 8000 BC near eastern farmers supports an early neolithic pioneer maritime colonization of Mainland Europe through Cyprus and the Aegean Islands." *PLoS Genetics, 2014, vol. 10, num. 6, p. e1004401* (2014)., Lacan, Marie, et al. "Ancient DNA reveals male diffusion through the Neolithic Mediterranean route." *Proceedings of the National Academy of Sciences* 108.24 (2011): 9788-9791. and Paschou, P, et al. "Maritime route of colonization of Europe." *Proceedings of the National Academy of Sciences* 111.25 (2014): 9211-9216.

519   Holinshed, Raphael, d. 1580?. *The Chronicles of England, Scotland and Ireland*. London *supra* The first inhabitation of Ireland, &c.:, http://goo.gl/OtCSgh and http://english.nsms.ox.ac.uk/holinshed/texts.php?text1=1587_0310 (Acessed on 09 April 2018)

520   Campion, Saint Edmund. *A Historie of Ireland, written in the Yeare 1571.* Hibernia Press, 1809.

521   Camden, William. *Britannia:(1607)* accessed at http://www.philological.bham.ac.uk/cambrit/scoteng.html#argyl1 on 28 August 2018.

522   See https://en.wikipedia.org/wiki/Argyll (Accessed on 28 August 2018)

523   McLauchlan, Thomas. «The Dean of Lismore›s Book.» (1862).

524   At page xxxii of the Introduction.

525   Northumberland was once part of an independent kingdom that stretched from Edinburgh, in Scotland, to the Humber River.

526   Keating, Geoffrey. "*The General History of Ireland, trans.*" *Dermod O'Connor (Dublin, 1723)* (1841)

527   Haverty, M. *The History of Ireland, Ancient and Modern: Derived from Our Native Annals, from the Most Recent Researches of Eminent Irish Scholars and Antiquaries, from the State Papers, and from All the Resources of Irish History Now Available.* J. Duffy, 1867.

528   "Anno Mundi (Latin: "in the year of the world"), abbreviated as AM or A.M., or Year After Creation, refers to a Calendar era based on the biblical creation of the world. Year zero AM is based on Rabbinic estimates for the year of creation; though various years have been suggested by various Jewish scholars, the generally accepted date of zero AM in Judaism corresponds to the Christian date of October 7th, 3761 BC." See http://en.wikipedia.org/wiki/Anno_Mundi

529   Referring to O'Flaherty, Roderic. "Ogygia, or, a chronological account of Irish events: collected from very ancient documents. Written originally in Latin by Roderic O'Flaherty, Esq. Translated by the Revd. James Hely, AB..." (1793) in which it was written that "*In the beginning of summer (2934 AM) on the kalends of May on the fifth day of the week and the seventh of the moon the Milesians, that is the eight sons of Golam, the Spanish soldier, with their relations and kinsmen, planted a Scots colony of Scythian origin in Ireland; which had been the fifth since the deluge ...*" Part III, page 29, Chapter XVI

530   Chapter 2, p. 13

531   Moore, T. *The history of Ireland.* Baudry, 1837.

532   Vol. 1, p. 62

533   See Wikipedia, Plato, http://en.wikipedia.org/wiki/Plato .

534   See notes on Appollodorus above

535   See http://en.wikipedia.org/wiki/Éire

536   See http://en.wikipedia.org/wiki/Éire

537   Schrijver, P. "Varia V. Non-Indo-European Surviving in Ireland in the First Millennium AD." Ériu (2000): 195-199.1

538   See also http://en.wikipedia.org/wiki/Goidelic_substrate_hypothesis (Accessed on 29 January 2015)

539   This was not unique to Ireland. It has been proposed that the current languages of Western Europe, including the Goidelic languages, have, over the millennia, been influenced to varying degrees by 'Vasconic' languages (the only remaining one being the Basque language) and Afroasiatic languages. Venneman proposes that Afroasiatics (or, as he refers to them, Semitidic peoples) may have colonised colonized coastal regions of Western and Northern Europe from as early as the fifth millennium BCE. See: Baldi, P, and B. R Page. "Europa Vasconica-Europa Semitica: Theo Vennemann, Gen. Nierfeld, in: Patrizia Noel Aziz Hanna (Ed.), Trends in Linguistics, Studies and Monographs 138, Mouton de Gruyter, Berlin, 2003, pp. xxii+ 977." *Lingua* 116.12 (2006): 2183-2220. And

Vennemann, T. "Atlantis Semitica: structural contact features in Celtic and English." *Amsterdam Studies In The Theory And History Of Linguistic Science Series 4* (1999): 351-370.

540   Referred to earlier.

541   Keating does not mention a name such as Gadeira but does mention Gaedheal glas ("Gathelus") and the Milesians or the sons of Míleadh as he calls them.

542   Edwards, Ceiridwen J., et al. *"Dual origins of dairy cattle farming – evidence from a comprehensive survey of European Y-chromosomal variation." PLoS One* 6.1 (2011): e15922.

543   *"Circumstantial evidence is evidence that relies on an inference to connect it to a conclusion of fact – like a fingerprint at the scene of a crime. ... On its own, circumstantial evidence allows for more than one explanation. Different pieces of circumstantial evidence may be required, so that each corroborates the conclusions drawn from the others. Together, they may more strongly support one particular inference over another."* See https://en.wikipedia.org/wiki/Circumstantial_evidence

544   See https://en.wikipedia.org/wiki/Straight-tusked_elephant and https://en.wikipedia.org/wiki/Woolly_mammoth .

545   See also Stuart, A. J. "The extinction of woolly mammoth (Mammuthus primigenius) and straight-tusked elephant (Palaeoloxodon antiquus) in Europe." *Quaternary International* 126 (2005): 171-177.

546   This is translated in the Perseus Tufts translation (at paragraph 114e) as *"It brought forth also in abundance all the timbers that a forest provides for the labours of carpenters; and of animals it produced a sufficiency, both of tame and wild."*

547   See https://en.wikipedia.org/wiki/Straight-tusked_elephant

548   See http://www.bbc.co.uk/nature/20678793 and Li, Ji, et al. "The latest straight-tusked elephants (Palaeoloxodon)?"Wild elephants" lived 3000 years ago in North China." *Quaternary International* 281 (2012): 84-88. (Abstract)

549   Masseti, M. "On the Pleistocene occurrence of Elephas (Palaeoloxodon) antiquus in the Tuscan Archipelago, northern Tyrrhenian Sea (Italy)." *Hystrix, the Italian Journal of Mammalogy* 5.1-2 (1994).

550   https://en.wikipedia.org/wiki/Würm_glaciation (Accessed on 4 April 2018)

551   Wiley-Blackwell. "Mammoths Survived In Britain Until 14,000 Years Ago, New Discovery Suggests." ScienceDaily. ScienceDaily, 18 June 2009. <www.sciencedaily.com/releases/2009/06/090617201758.htm> See also Lister, Adrian M. "Late glacial mammoth skeletons (Mammuthus primigenius) from Condover (Shropshire, UK): anatomy, pathology, taphonomy and chronological significance." *Geological Journal* 44.4 (2009): 447-479., and Scourse, J. D., et al. "Late glacial remains of woolly mammoth (Mammuthus primigenius) from Shropshire, UK: stratigraphy, sedimentology and geochronology of the Condover site." *Geological Journal* 44.4 (2009): 392-413.

552   Ukkonen, P., et al. "Woolly mammoth (Mammuthus primigenius Blum.) and its environment in northern Europe during the last glaciation." *Quaternary Science Reviews* 30.5-6 (2011): 693-712.

553   Markova, A. K., et al. "New data on changes in the European distribution of the mammoth and the woolly rhinoceros during the second half of the Late Pleistocene and the early Holocene." *Quaternary International* 292 (2013): 4-14.

554   Guthrie, R. D. "Radiocarbon evidence of mid-Holocene mammoths stranded on an Alaskan Bering Sea island." *Nature* 429.6993 (2004): 746-749.

555   Vartanyan, S. L., et al. "Radiocarbon dating evidence for mammoths on Wrangel Island, Arctic Ocean, until 2000 BC." *Radiocarbon* 37.1 (1995): 1-6.

556   See also Martin, P. S. and A.J. Stuart. "Mammoth extinction: two continents and Wrangel Island." Radiocarbon, Vol. 37.1, 1995, 7-10 and Arslanov, Kh A., et al. "Consensus dating of mammoth remains from Wrangel Island." *Radiocarbon* 40.1 (1998): 289-294.

557   Van der Plicht, J., et al. "New Holocene refugia of giant deer (Megaloceros giganteus Blum.) in Siberia: updated extinction patterns." *Quaternary Science Reviews* 114 (2015): 182-188.

558   http://www.mineralsireland.ie/Mining+in+Ireland/ (Accessed 21 September 2014)

559   http://www.mineralsireland.ie/MiningInIreland (Accessed 5 April 2018)

560   "Brass is an alloy of copper and zinc; the proportions of zinc and copper can be varied to create a range of brasses with varying properties." See http://en.wikipedia.org/wiki/Brass (Accessed 21 September 2014)

561   See http://en.wikipedia.org/wiki/Brass (Accessed 21 September 2014)

562   http://www.libraryireland.com/SocialHistoryAncientIreland/III-XX-1.php (Accessed 21 September 2014)

563   http://en.wikipedia.org/wiki/Mining_in_Cornwall_and_Devon (Accessed 21 September 2014)

564   http://en.wikipedia.org/wiki/Mining_in_Cornwall_and_Devon (Accessed 21 September 2014)

565   *"Later writers — Posidonius, Diodorus Siculus, Strabo and others – call them smallish islands off ("some way off," Strabo says) the northwest coast of the Iberian Peninsula, which contained tin mines or, according to Strabo, tin and lead mines."* http://en.wikipedia.org/wiki/Cassiterides (Accessed 21 September 2014)

566   A Greek historian who wrote works of history between 60 and 30 BC. See http://en.wikipedia.org/wiki/Diodorus_Siculus (Accessed 21 September 2014)

567   *The Library of History (Book V) of Diodorus Siculus*, published in Vol. III of the Loeb Classical Library edition, 1939

568   *Timaeus*

569   *"Tyrrhenia* [which was also known as Etruria] ... *was a region of Central Italy, located in an area that covered part of what now are Tuscany, Lazio, and Umbria"*. See http://en.wikipedia.org/wiki/Etruria (Accessed 21 September 2014)

570   The upper palaeolithic covers the end of the Pleistocene era and is generally regarded as the period roughly coinciding with the appearance of behavioural modernity and before the advent of agriculture (See http://en.wikipedia.org/wiki/Upper_Paleolithic - accessed 21 September 2014)

571   The upper palaeolithic extended, very broadly speaking, from 50 000 BCE to 10 000 BCE and the Holocene from 10 000 BCE onwards.

572   González, A. M., et al. "The mitochondrial lineage U8a reveals a Paleolithic settlement in the Basque country." *BMC genomics* 7.1 (2006): 124.

573   See also Behar, D. M., et al. "The Basque paradigm: genetic evidence of a maternal continuity in the Franco-Cantabrian region since pre-Neolithic times." *The American Journal of Human Genetics* 90.3 (2012): 486-493.

574   See also Alonso, S., et al. "The place of the Basques in the European Y-chromosome diversity landscape." *European journal of human genetics* 13.12 (2005): 1293-1302.

575   Hill, E. W., M. A. Jobling, and D. G. Bradley. "Y-chromosome variation and Irish origins." *Nature* 404.6776 (2000): 351-352.

576   See Grindon, A. J., and A. Davison. "Irish Cepaea nemoralis land snails have a cryptic Franco-Iberian origin that is most easily explained by the movements of Mesolithic humans." *PloS one* 8.6 (2013): e65792.

577   Rosser, Z. H., et al. "Y-chromosomal diversity in Europe is clinal and influenced primarily by geography, rather than by language." *The American Journal of Human Genetics* 67.6 (2000): 1526-1543.

578   See 'Briefe relation of Ireland, and the diversity of Irish in the same [and] Priests in Ireland and Gentlemen gone abroad'. CELT: Corpus of Electronic Texts, http://www.ucc.ie/celt/published/T100077.html (Accessed 21 September 2014)

579   Torroni, A., et al. "A signal, from human mtDNA, of postglacial recolonization in Europe." *The American Journal of Human Genetics* 69.4 (2001): 844-852.

580   Tambets, K., et al. "The western and eastern roots of the Saami—the story of genetic "outliers" told by mitochondrial DNA and Y chromosomes." *The American Journal of Human Genetics* 74.4 (2004): 661-682.

581   Achilli, A., et al. "Saami and Berbers – an unexpected mitochondrial DNA link." *The American Journal of Human Genetics* 76.5 (2005): 883-886.

582   See Stewart, J. R., et al. "Refugia revisited: individualistic responses of species in space and time." *Proceedings of the Royal Society B: Biological Sciences* 277.1682 (2010): 661-671, Figure 1 AT P. 663

583   Whilst probably not corroboration of this suggestion it might be worth noting that it is stated in *The Roman History of Ammianus Marcellinus*, Book XV, Chapt. 9, [translated by Rolfe, J. C. 1935–39. Ammianus Marcellinus. Loeb Classical Library. 3 vols. Cambridge, Mass.] that, according to the Druids, " *... a part of the people* [of Gaul] *was in fact indigenous, but that others also poured in from the remote islands and the regions across the Rhine, driven from their homes by continual wars and by the inundation of the stormy sea.*"

584   Frank, R. M. "Recovering European ritual bear hunts: a comparative study of Basque and Sardinian ursine carnival performances." *INSULA: Quaderno di cultura sarda* 3 (2008): 41-97.

585   Leschber, C. "Latin tree names and the European substratum." *Studia Linguistica Universitatis Iagellonicae Cracoviensis* 129 (2012): 117-125.

                                                ATLANTIS, FOUND?

586 Haarmann, H. "Indo-Europeanization—the seven dimensions in the study of a never ending process." *Documenta Praehistorica XXXIV. Neolithic Studies* 14 (2007): 155-175.

587 See Vennemann, T. «Atlantis Semitica: structural contact features in Celtic and English.» *Amsterdam Studies In The Theory And History Of Linguistic Science Series 4* (1999): 351-370 and Baldi, P, and B. R Page. «Europa Vasconica-Europa Semitica: Theo Vennemann, Gen. Nierfeld, in: Patrizia Noel Aziz Hanna (Ed.), Trends in Linguistics, Studies and Monographs 138, Mouton de Gruyter, Berlin, 2003, pp. xxii+ 977.» *Lingua* 116.12 (2006): 2183-2220.

588 Leschber, C. "Latin tree names and the European substratum." *Studia Linguistica Universitatis Iagellonicae Cracoviensis* 129 (2012): 117-125.

589 See also Section 8.2.2.4

590 Boutkan, D, and M G. Kossmann. "Some Berber parallels of European substratum words." *Journal of Indo-European Studies* 27.1/2 (1999): 87 as cited at page 118 in Leschber, Corinna. "Latin tree names and the European substratum." *Studia Linguistica Universitatis Iagellonicae Cracoviensis* 129 (2012): 117-125.

591 i.e. indigenous Asian, European, Australian and American populations.

592 Fregel, R., et al. "The maternal aborigine colonization of La Palma (Canary Islands)." *European Journal of Human Genetics* 17.10 (2009): 1314-1324.

593 Maca-Meyer, N., et al. "Ancient mtDNA analysis and the origin of the Guanches." *European journal of human genetics* 12.2 (2004): 155-162.

594 Fu, Qiaomei, et al. "The genetic history of Ice Age Europe." *Nature* (2016)

595 i.e. after ~14 000 years ago or after ~12 000 BCE.

596 *Critias*

597 *Timaeus*

598 Barber, E. W. and P. T. Barber. *When they severed earth from sky: how the human mind shapes myth.* Princeton University Press, 2006.

599 See http://en.wikipedia.org/wiki/8.2_kiloyear_event

600 Accepting that the Holocene commenced at 10 000 BCE

601 https://platoproject.gr/mom-1/

602 http://www.perseus.tufts.edu/hopper/text?doc=Perseus:abo:tlg,0059,031:25d (Accessed on 20 April 2018)

603 Approximately 12 500 years ago

604 Approximately 10 500 years ago

605 Although I shall focus on this time period it must be understood that my proposition is that Hatton Rockall sank over millennia and that this does not exclude the possibility that there may have been subsequent cataclysmic events (I think, in particular, of the so-called "8.2 kiloyear event" (See http://en.wikipedia.org/wiki/8.2_kiloyear_event) when there was further massive flooding.) which hastened, or even slowed, the sinking of Hatton Rockall.

606 See Wikipedia at http://en.wikipedia.org/wiki/List_of_prehistoric_lakes for a list and details of various prehistoric lakes.

607 See Wikipedia at http://en.wikipedia.org/wiki/Deluge_(prehistoric)

608 See also Meinsen, J. et al. "Middle Pleistocene (Saalian) lake outburst floods in the Münsterland Embayment (NW Germany): impacts and magnitudes."

Quaternary Science Reviews 30.19 (2011): 2597-2625., Komatsu, G. et al. "Quaternary paleolake formation and cataclysmic flooding along the upper Yenisei River." *Geomorphology* 104.3 (2009): 143-164., and Kong, P. et al. "Moraine dam related to late Quaternary glaciation in the Yulong Mountains, southwest China, and impacts on the Jinsha River." *Quaternary Science Reviews* 28.27 (2009): 3224-3235.

609   See also Jones, S. *The Serpent's Promise: The Bible Retold as Science*. Hachette UK, 2013. at page 221 *et seq.*

610   McMillan, M. et al. "Three-dimensional mapping by CryoSat-2 of subglacial lake volume changes." *Geophysical Research Letters* 40.16 (2013): 4321-4327.

611   Antarctic lake released massive under ice flood at http://www.abc.net.au/science/articles/2013/07/04/3796096.htm (Accessed on 19 September 2014)

612   Tang, Liujuan, et al. «Direct energy estimation of the 2011 Japan tsunami using deep-ocean pressure measurements.» *Journal of Geophysical Research: Oceans* 117.C8 (2012).

613   Approximately 12 000 years ago

614   Keigwin, L. D. et al. "Rapid sea-level rise and Holocene climate in the Chukchi Sea." *Geology* 34.10 (2006): 861-864. Meyer, H., et al. "Permafrost evidence for severe winter cooling during the Younger Dryas in northern Alaska." *Geophysical Research Letters* 37.3 (2010).

615   The Fram Strait is the main gateway for the exchange of water between the Arctic and Atlantic oceans. See also:
Zamelczyk, K., et al. "Paleoceanographic changes and calcium carbonate dissolution in the central Fram Strait during the last 20ka." *Quaternary Research* 78.3 (2012): 405-416.

616   Even if the Bering Strait were open, it is substantially narrower and shallower than the Norwegian Sea and, as such, flood waters entering the Arctic Ocean water would exit in proportionately greater volumes through the Norwegian Sea than through the Bering Strait.

617   Based upon a map obtained from http://maps.ngdc.noaa.gov/viewers/bathymetry/ (Accessed on 22 October 2014) The original map is United States Government material and is not subject to copyright protection within the United States and is in the public domain.

618   L. D. Keigwin, S. Klotsko, N. Zhao, B. Reilly, L. Giosan, N. W. Driscoll. Deglacial floods in the Beaufort Sea preceded Younger Dryas cooling. *Nature Geoscience*, 2018

619   I do not consider to be relevant any floods which emptied in such a direction that none of the flood waters would have had a direct effect on the Hatton Rockall area – such as the Altai flood which flooded into the Aral Sea, the Caspian Sea and beyond. See http://en.wikipedia.org/wiki/Altai_flood (Accessed on 29 January 2015)

620   See https://en.wikipedia.org/wiki/Lake_Agassiz (Accessed on 4 September 2018)

621   Mann, J. D., et al. "The volume and paleobathymetry of glacial Lake Agassiz." *Journal of Paleolimnology* 22.1 (1999): 71-80.

622   11.5 – 11.0 ^{14}C ka BP

623   9.9–9.5 ^{14}C ka, BP

624   See https://en.wikipedia.org/wiki/List_of_lakes_by_volume (Accessed on 16 August 2018)

625   Table 1 in Teller, J. T., D. W. Leverington, and J. D. Mann. "Freshwater outbursts to the oceans from glacial Lake Agassiz and their role in climate change during the last deglaciation." *Quaternary Science Reviews* 21.8 (2002): 879-887.

626   See also Leverington, D. W., J. D. Mann, and J. T. Teller. "Changes in the bathymetry and volume of glacial Lake Agassiz between 11,000 and 9300 ^{14}C yr BP." *Quaternary Research* 54.2 (2000): 174-181. (Also Leverington, David W., Jason D. Mann, and James T. Teller. "Changes in the bathymetry and volume of glacial Lake Agassiz between 9200 and 7700 ^{14}C yr BP." *Quaternary Research* 57.2 (2002): 244-252.)

627   Teller, J. T. and D. W. Leverington. "Outbursts from Lake Agassiz and their possible impact on coastal environments." *Conference abstracts: Environmental Catastrophes and Recoveries in the Holocene. Uxbridge, UK: Brunel University.* 2002.

628   See also Murton, J. B., et al. "Identification of Younger Dryas outburst flood path from Lake Agassiz to the Arctic Ocean." *Nature* 464.7289 (2010): 740-743.

629   Keigwin, L.D., S. Klotsko, N. Zhao, B. Reilly, L. Giosan, N. W. Driscoll. Deglacial floods in the Beaufort Sea preceded Younger Dryas cooling. *Nature Geoscience*, 2018

630   Murton, J. B., et al. "Identification of Younger Dryas outburst flood path from Lake Agassiz to the Arctic Ocean." *Nature* 464.7289 (2010): 740-743.

631   Described as "two major *high energy fluvial episodes*"

632   See Figure 9.1 for the approximate position of the current mouth

633   13 000 (+/- 200) years and 11 700 (+/- 100) ^{14}C years BP

634   See also Condron, A. and P. Winsor. "Meltwater routing and the Younger Dryas." *Proceedings of the National Academy of Sciences* 109.49 (2012): 19928-19933.

635   http://en.wikipedia.org/wiki/Younger_Dryas

636   11 700 (+/- 100) years and 9 300 (+/- 700) years BP

637   Fisher, T. G., D. G. Smith, and J. T. Andrews. "Preboreal oscillation caused by a glacial Lake Agassiz flood." *Quaternary Science Reviews* 21.8 (2002): 873-878.

638   11 300 BP

639   See also Björck, S., et al. "The Preboreal oscillation around the Nordic Seas: terrestrial and lacustrine responses." *Journal of Quaternary Science* 12.6 (1997): 455-465.

640   Margold, M. et al. "Glacial Lake Vitim, a 3 000km³ outburst flood from Siberia to the Arctic Ocean." *Quaternary Research* 76.3 (2011): 393-396.

641   13 000 BP

642   Lioubimtseva, E. U., S. P. Gorshkov, and J. M. Adams. "A giant Siberian lake during the last glacial: evidence and implications." *Unpublished paper. http://*

www. esd. ornl. gov/projects/qen/lake. Html (2001). (Accessed on 08 October 2014)

643   Reuther, A. U., et al. "Constraining the timing of the most recent cataclysmic flood event from ice-dammed lakes in the Russian Altai Mountains, Siberia, using cosmogenic in situ 10Be." *Geology* 34.11 (2006): 913-916.

644   There were also, it would appear, various late glacial floods in Swedish Lapland. These caused, amongst others devastation, the erosion of hills and 'boulder deltas'. See Elfstrom, A. "The Båldakatj Boulder Delta, Lapland, Northern Sweden." Geografiska Annaler. Series A, Physical Geography, Vol. 65, No. 3/4 (1983), pp. 201-225 and Elfström, Å. and L. Rossbacher "Erosional Remnants in the Båldakatj Area, Lapland, Northern Sweden. A Terrestrial Analog for Martian Landforms" Geografiska Annaler. Series A, Physical Geography, Vol. 67, No. 3/4 (1985), pp. 167-176

645   At about 11 500 years ^{14}C BP, i.e. approximately 9 500 BCE.

646   Jakobsson, M. et al. "Reconstructing the Younger Dryas ice dammed lake in the Baltic Basin: Bathymetry, area and volume." *Global and Planetary Change* 57.3 (2007): 355-370.

647   Teller, J. T., and D. W. Leverington. "Outbursts from Lake Agassiz and their possible impact on coastal environments." *Conference abstracts: Environmental Catastrophes and Recoveries in the Holocene. Uxbridge, UK: Brunel University.* 2002.

648   Fisher, T. G., D. G. Smith, and J. T. Andrews. "Preboreal oscillation caused by a glacial Lake Agassiz flood." *Quaternary Science Reviews* 21.8 (2002): 873-878.

649   Murton, Julian B., et al. "Identification of Younger Dryas outburst flood path from Lake Agassiz to the Arctic Ocean." *Nature* 464.7289 (2010): 740-743

650   Lohne, Ø. S., et al. "Calendar year age estimates of Allerød–Younger Dryas sea-level oscillations at Os, western Norway." *Journal of Quaternary Science* 19.5 (2004): 443-464.

651   Keigwin, L. D., S. Klotsko, N. Zhao, B. Reilly, L. Giosan, N. W. Driscoll. Deglacial floods in the Beaufort Sea preceded Younger Dryas cooling. *Nature Geoscience*, 2018

652   See, amongst many others, Gupta, H. K. "A review of recent studies of triggered earthquakes by artificial water reservoirs with special emphasis on earthquakes in Koyna, India." *Earth-Science Reviews* 58.3 (2002): 279-310., Talwani, P. "On the nature of reservoir-induced seismicity." *Seismicity Associated with Mines, Reservoirs and Fluid Injections.* Birkhäuser Basel, 1998. 473-492., Durá-Gómez, I. and P. Talwani. "Reservoir-induced seismicity associated with the Itoiz Reservoir, Spain: a case study." *Geophysical Journal International* 181.1 (2010): 343-356., Dai, M.. *et al.* "A Study on the Relationship between Water Levels and Seismic Activity in the Three Gorges Reservoir." *Institute of Seismology, China Earthquake Administration, Wuhan (tr. and pub in English by Probe International 2011), http://probeinternational. org/ library/wp—content/uploads/ZOI1/06/3—Gorges—Report—26-5. pdf* (2010).

653   Hainzl, S., et al. "Evidence for rainfall-triggered earthquake activity." *Geophysical Research Letters* 33.19 (2006)., Jimenez, M-J., and M. Garcia-

Fernández. "Occurrence of shallow earthquakes following periods of intense rainfall in Tenerife, Canary Islands." *Journal of volcanology and geothermal research* 103.1 (2000): 463-468.

654   Talwani, P. "On the nature of reservoir-induced seismicity." *Pure and applied geophysics* 150.3-4 (1997): 473-492.

655   Utkucu, M. "Implications for the water level change triggered moderate (M≥ 4.0) earthquakes in Lake Van basin, Eastern Turkey." *Journal of seismology* 10.1 (2006): 105-117.

656   See also Brandes, C., U. Polom, and J. Winsemann. "Reactivation of basement faults: interplay of ice-sheet advance, glacial lake formation and sediment loading." *Basin Research* 23.1 (2011): 53-64.

657   See Table 1 in Chen, Y. "Did the reservoir impoundment trigger the Wenchuan earthquake?." *Science in China Series D: Earth Sciences* 52.4 (2009): 431-433.

658   See https://en.wikipedia.org/wiki/List_of_reservoirs_by_volume (Accessed on 16 August 2018)

659   See https://en.wikipedia.org/wiki/Lake_Kariba (Accessed on 04-09-18)

660   See https://earthobservatory.nasa.gov/images/87485/the-decline-of-lake-kariba Accessed on 31-08-18

661   See https://www.herald.co.zw/quake-rattles-kariba/ (Accessed on 31-08-18)

662   Brothers, D. S., K. M. Luttrell, and J. D. Chaytor. "Sea-level–induced seismicity and submarine landslide occurrence." *Geology* 41.9 (2013): 979-982.

663   Hampel, A. et al. "Three-dimensional numerical modelling of slip rate variations on normal and thrust fault arrays during ice cap growth and melting." *Journal of Geophysical Research: Solid Earth (1978–2012)* 114.B8 (2009).

664   See also Hampel, A., R. Hetzel, and G. Maniatis. "Response of faults to climate-driven changes in ice and water volumes on Earth's surface." *Philosophical Transactions of the Royal Society A: Mathematical, Physical and Engineering Sciences* 368.1919 (2010): 2501-2517.

665   Hampel, A., R. Hetzel, and G. Maniatis. "Response of faults to climate-driven changes in ice and water volumes on Earth's surface." *Philosophical Transactions of the Royal Society A: Mathematical, Physical and Engineering Sciences* 368.1919 (2010): 2501-2517.

666   Fennoscandia is the region comprising the Scandinavian Peninsula, Finland, Karelia, and the Kola Peninsula. (See http://en.wikipedia.org/wiki/ Fennoscandia accessed on 19 September 2014)

667   Arvidsson, R. "Fennoscandian earthquakes: whole crustal rupturing related to postglacial rebound." *Science* 274.5288 (1996): 744-746.

668   *M*w refers to the 'moment magnitude scale' which is used by seismologists to measure the size of earthquakes in terms of the energy released and is different from the older Richter Scale. See https://en.wikipedia.org/wiki/Moment_ magnitude_scale

669   Hampel, A., R. Hetzel, and G. Maniatis. "Response of faults to climate-driven changes in ice and water volumes on Earth's surface." *Philosophical Transactions*

of the Royal Society A: Mathematical, Physical and Engineering Sciences 368.1919 (2010): 2501-2517.

670   Lagerbäck, R. and M. Sundh. *Early Holocene faulting and paleoseismicity in northern Sweden*. Na, 2008., Geological survey of Sweden. Research paper, C 836. and Baeckstroem, A., N. Rantakokko, and M. V. Ask. "Structural Analysis of the Pärvie Fault in Northern Scandinavia." *AGU Fall Meeting Abstracts*. Vol. 1. 2011.

671   Baeckstroem, A., N. Rantakokko, and M. V. Ask. "Structural Analysis of the Pärvie Fault in Northern Scandinavia." *AGU Fall Meeting Abstracts*. Vol. 1. 2011 and Arvidsson, Ronald. "Fennoscandian earthquakes: whole crustal rupturing related to postglacial rebound." *Science* 274.5288 (1996): 744-746.

672   See also Olesen, O. et al. "Neotectonic deformation in Norway and its implications: a review." *NORSK GEOLOGISK TIDSSKRIFT* 84.1 (2004): 3-34., and Lagerbäck, R. and M. Sundh. *Early Holocene faulting and paleoseismicity in northern Sweden*. Na, 2008., Geological survey of Sweden. Research paper, C 836.

673   11 500 ^{14}C BP and 9 500 ^{14}C BP

674   See also Sutinen, R., M. Piekkari and M. Middleton. "Glacial geomorphology in Utsjoki, Finnish Lapland proposes Younger Dryas fault-instability." *Global and Planetary Change* 69.1 (2009): 16-28.

675   Dehls, J. F. et al. "Neotectonic faulting in northern Norway; the Stuoragurra and Nordmannvikdalen postglacial faults." *Quaternary Science Reviews* 19.14 (2000): 1447-1460., Olesen, O., Blikra, L.H., Braathen, A., Dehls, J.F., Olsen, L., Rise, L., Roberts, D., Riis, F., Faleide, J.I. & Anda, E.: Neotectonic deformation in Norway and its implications: a review. Norwegian Journal of Geology, Vol. 84, pp. 3-34

676   Dehls, J. F., et al. "Neotectonic faulting in northern Norway; the Stuoragurra and Nordmannvikdalen postglacial faults." *Quaternary Science Reviews* 19.14 (2000): 1447-1460.,

677   11 000^{14}C years ago

678   10 000 and 9 500^{14}C years ago

679   Which had a magnitude of 9.0–9.1 (M_w).

680   See http://en.wikipedia.org/wiki/2011_T%C5%8Dhoku_earthquake_and_tsunami (Accessed on 8 March 2018)

681   Sutinen, R., M. Piekkari, and M. Middleton. "Glacial geomorphology in Utsjoki, Finnish Lapland proposes Younger Dryas fault-instability." *Global and Planetary Change* 69.1 (2009): 16-28.

682   See also Sutinen, R. et al. "Airborne LiDAR detection of postglacial faults and Pulju moraine in Palojärvi, Finnish Lapland." *Global and Planetary Change* 115 (2014): 24-32., and Kotilainen, A. and K. Hutri. "Submarine Holocene sedimentary disturbances in the Olkiluoto area of the Gulf of Bothnia, Baltic Sea: a case of postglacial palaeoseismicity." *Quaternary Science Reviews* 23.9 (2004): 1125-1135., Mörner, N-A. "Liquefaction and varve deformation as evidence of paleoseismic events and tsunamis. The autumn 10,430 BP case in Sweden." *Quaternary Science Reviews* 15.8 (1996): 939-948.

683   Karlin, R. E. and S. E. B. Abella. "Paleoearthquakes in the Puget Sound region recorded in sediments from Lake Washington, USA." *Science* 258.5088 (1992): 1617-1620., Bacon, S. N. and S. K. Pezzopane. "A 25,000-year record of earthquakes on the Owens Valley fault near Lone Pine, California: Implications for recurrence intervals, slip rates, and segmentation models." *Geological Society of America Bulletin* 119.7-8 (2007): 823-847., Adams, J. "Postglacial faulting in eastern Canada: nature, origin and seismic hazard implications." *Tectonophysics* 163.3 (1989): 323-331., Briggs, R. W. and S. G. Wesnousky. "Late Pleistocene and Holocene paleoearthquake activity of the Olinghouse fault zone, Nevada." *Bulletin of the Seismological Society of America* 95.4 (2005): 1301-1313.

684   Takada, Y. and Y. Fukushima. "Volcanic subsidence triggered by the 2011 Tohoku earthquake in Japan." *Nature Geoscience* 6.8 (2013): 637-641. and Pritchard, M. E., et al. "Subsidence at southern Andes volcanoes induced by the 2010 Maule, Chile earthquake." *Nature Geoscience* 6.8 (2013): 632-636.

685   Takada, Y. and Y. Fukushima. "Volcanic subsidence triggered by the 2011 Tohoku earthquake in Japan." *Nature Geoscience* 6.8 (2013): 637-641

686   Walter, T. R. "How a tectonic earthquake may wake up volcanoes: Stress transfer during the 1996 earthquake–eruption sequence at the Karymsky Volcanic Group, Kamchatka." *Earth and Planetary Science Letters* 264.3 (2007): 347-359.

687   Marzocchi, W. "Remote seismic influence on large explosive eruptions." *Journal of Geophysical Research: Solid Earth (1978–2012)* 107.B1 (2002): EPM-6., pages EPM 6-1–EPM 6-7, January 2002.

688   See also Eggert, S. and T. R. Walter. "Volcanic activity before and after large tectonic earthquakes: observations and statistical significance." *Tectonophysics* 471.1 (2009): 14-26.

689   Husen, S., et al. "Changes in geyser eruption behavior and remotely triggered seismicity in Yellowstone National Park produced by the 2002 M 7.9 Denali fault earthquake, Alaska." *Geology* 32.6 (2004): 537-540.

690   See http://en.wikipedia.org/wiki/1960_Valdivia_earthquake (Accessed on 27 October 2014)

691   Watt, S. F. L., D. M. Pyle, and T. A. Mather. "The influence of great earthquakes on volcanic eruption rate along the Chilean subduction zone." *Earth and Planetary Science Letters* 277.3 (2009): 399-407.

692   See also Harris, A. J. L, and M. Ripepe. "Regional earthquake as a trigger for enhanced volcanic activity: evidence from MODIS thermal data." *Geophysical Research Letters* 34.2 (2007)., Walter, T. R. "How a tectonic earthquake may wake up volcanoes: Stress transfer during the 1996 earthquake–eruption sequence at the Karymsky Volcanic Group, Kamchatka." *Earth and Planetary Science Letters* 264.3 (2007): 347-359. and Devastating Long-Distance Impact of Earthquakes, Universität Bonn (2013, July 23). Devastating long-distance impact of earthquakes. ScienceDaily. Retrieved July 24, 2013, from http://www.sciencedaily.com/releases /2013/07/130723073957.htm and Cembrano, J. and L. Lara. "The link between volcanism and tectonics in the southern

volcanic zone of the Chilean Andes: A review." *Tectonophysics* 471.1 (2009): 96-113.

693   Manga, M. and E. Brodsky. "Seismic triggering of eruptions in the far field: volcanoes and geysers." *Annu. Rev. Earth Planet. Sci* 34 (2006): 263-291.

694   See also Delle Donne, D. et al. "Earthquake-induced thermal anomalies at active volcanoes." *Geology* 38.9 (2010): 771-774.

695   Wauthier, C., B. Smets, and D. Keir (2015), Diking-induced moderate-magnitude earthquakes on a youthful rift border fault: The 2002 Nyiragongo-Kalehe sequence, D.R. Congo, Geochem. Geophys. Geosyst., 16, 4280– 4291

696   Zielinski, Gregory A., et al. "A 110,000-yr record of explosive volcanism from the GISP2 (Greenland) ice core." *Quaternary Research* 45.2 (1996): 109-118.

697   i.e. 6 000 and 17 000 years ago

698   i.e. 7 000 and 13 000 years ago

699   Zielinski, Gregory A. "Use of paleo-records in determining variability within the volcanism–climate system." *Quaternary Science Reviews* 19.1 (2000): 417-438.

700   University of Cambridge. "Increase in volcanic eruptions at the end of the ice age caused by melting ice caps and erosion." ScienceDaily. ScienceDaily, 1 February 2016. <www.sciencedaily.com/releases/2016/02/160201141731.htm>

701   *Sternai, P., L. Caricchi, S. Castelltort, and J.-D. Champagnac (2016), Deglaciation and glacial erosion: a joint control on magma productivity by continental unloading, Geophys. Res. Lett., 43, doi:10.1002/2015GL067285. (Abstract)*

702   Zielinski, Gregory A., et al. "A 110,000-yr record of explosive volcanism from the GISP2 (Greenland) ice core." *Quaternary Research* 45.2 (1996): 109-118.

703   See Wikipedia at http://en.wikipedia.org/wiki/1815_eruption_of_Mount_Tambora and http://en.wikipedia.org/wiki/Year_Without_a_Summer accessed on 19 September 2014.

704   See paragraph 3.2.2.3

705   Pfeffer, Julia, et al. "Decoding the origins of vertical land motions observed today at coasts." *Geophysical Journal International* 210.1 (2017): 148-165 at p. 149

706   Frederikse, Thomas, Riccardo EM Riva, and Matt A. King. "Ocean Bottom Deformation Due To Present-Day Mass Redistribution and Its Impact on Sea Level Observations." *Geophysical Research Letters* 44.24 (2017).

707   Frederikse, *et al supra* at p. 7

708   Thousands of square kilometres

709   See for example Zhang, Kui, et al. «Monitoring ground surface deformation over the North China Plain using coherent ALOS PALSAR differential interferograms.» *Journal of Geodesy* 87.3 (2013): 253-265.

710   Poland, J. F. "Land subsidence in the San Joaquin Valley, California, and its effect on estimates of ground-water resources." *Internat. Assoc. of Sci. Hydrology, Publ* 52 (1960): 324-335.

711   Ireland, R. L., J. F. Poland and F. S. Riley. *Land subsidence in the San Joaquin Valley, California, as of 1980.* US Government Printing Office, 1984.

712   Farr, T. G, C. Jones and Z. Liu, "Progress Report: Subsidence in the Central Valley California " Jet Propulsion Laboratory, California Institute of Technology, 2015.

713   Kim, Jin-Woo, and Zhong Lu. "Association between localized geohazards in West Texas and human activities, recognized by Sentinel-1A/B satellite radar imagery." *Scientific reports* 8.1 (2018): 4727

714   Southern Methodist University. "Radar images show large swath of Texas oil patch is heaving and sinking at alarming rates." ScienceDaily. ScienceDaily,21 March 2018. <www.sciencedaily.com/releases/2018/03/180321110855.htm>

715   https://www.sciencealert.com/a-mysterious-seismic-wave-rumbled-earth-and-scientists-can-t-explain-it

716   https://www.clicanoo.re/Societe/Article/2018/12/01/Activite-sismique-Mayotte-lhypothese-volcanique-se-precise_559712 (Accessed on 3 December 2018)

717   See paragraph 6.4.3 below

718   Barletta, Valentina R., et al. «Observed rapid bedrock uplift in Amundsen Sea Embayment promotes ice-sheet stability.» *Science* 360.6395 (2018): 1335-1339.

719   *M.J. Hoggard et al. 'Global dynamic topography observations reveal limited influence of large-scale mantle flow.' Nature Geoscience (2016). DOI: 10.1038/ ngeo2709*

720   University of Cambridge. "Map of flow within the Earth's mantle finds the surface moving up and down 'like a yo-yo'." ScienceDaily. ScienceDaily, 9 May 2016. <www.sciencedaily.com/releases/2016/05/160509115116.htm> .

721   See http://gizmodo.com/we-were-totally-wrong-about-whats-happening-inside-eart-1775495644

722   Massachusetts Institute of Technology. "Mystery of India's rapid move toward Eurasia 80 million years ago explained." ScienceDaily. ScienceDaily, 4 May 2015. <www.sciencedaily.com/releases/2015/05 /150504120815.htm>. See also Jagoutz, Oliver, et al. "Anomalously fast convergence of India and Eurasia caused by double subduction." *Nature Geoscience* 8.6 (2015): 475-478. and van Hinsbergen, Douwe JJ, et al. "Acceleration and deceleration of India-Asia convergence since the Cretaceous: Roles of mantle plumes and continental collision." *Journal of Geophysical Research: Solid Earth (1978–2012)* 116.B6 (2011).

723   Aitchison, J. C., J. R. Ali, and A. M. Davis. "When and where did India and Asia collide?." *Journal of Geophysical Research: Solid Earth (1978–2012)* 112.B5 (2007). at par. 16

724   During the time between, approximately, 80 and 50 million years ago.

725   See http://www.ga.gov.au/scientific-topics/positioning-navigation/datum-modernisation and https://www.sciencealert.com/australia-s-about-to-move-1-5-metres-to-the-north (Both accessed on 28 September 2018)

726   Darwin, C. "A naturalist's voyage round the world: Journal of researches into the natural history and geology of the countries visited during the voyage round the world of HMS Beagle under the command of Captain Fitz Roy." *RN* (1913) at p. 12.

727   Japsen, P. et al. "Episodic burial and exhumation in NE Brazil after opening of the South Atlantic." *Geological Society of America Bulletin* 124.5-6 (2012): 800-816.

728   A 'passive continental margin' is defined by Wikipedia as the " *transition between oceanic and continental lithosphere that is not an active plate margin*". See Wikipedia at http://en.wikipedia.org/wiki/Passive_margin , (Accessed on 3 April 2018.)

729   For further discussion on volcanic passive margins see Wikipedia, http://en.wikipedia.org/wiki/Volcanic_passive_margin (Accessed on 20 September 2014), and "Volcanic passive margins." by L Geoffroy. "Volcanic passive margins." *Comptes Rendus Geoscience* 337.16 (2005): 1395-1408

730   Japsen, P. et al. "Elevated, passive continental margins: Not rift shoulders, but expressions of episodic, post-rift burial and exhumation." *Global and Planetary Change* 90 (2012): 73-86.

731   See also Wikipedia at http://en.wikipedia.org/wiki/Passive_margin, Doré, A. G., et al. "Principal tectonic events in the evolution of the northwest European Atlantic margin." *Geological Society, London, Petroleum Geology Conference series*. Vol. 5., p. 41-61, Geological Society of London, 1999.

732   Japsen, P. et al. "Elevated, passive continental margins: Not rift shoulders, but expressions of episodic, post-rift burial and exhumation." *Global and Planetary Change* 90 (2012): 73-86. and Japsen, P., et al. "Episodic uplift and exhumation along North Atlantic passive margins: implications for hydrocarbon prospectivity." *Geological Society, London, Petroleum Geology Conference series*. Vol. 7. Geological Society of London, 2010. p. 979-1004

733   See also Bonow, J. M. et al. "Elevated erosion surfaces in central West Greenland and southern Norway: their significance in integrated studies of passive margin development." *Norwegian Journal of Geology* 87 (2007): 197-206.

734   Green, P. F., et al. "Thermochronology, erosion surfaces and missing section in West Greenland." *Journal of the Geological Society* 168.4 (2011): 817-830.

735   See also Holford, S. P. et al. "Regional intraplate exhumation episodes related to plate-boundary deformation." *Geological Society of America Bulletin* 121.11-12 (2009): 1611-1628.

736   Hartley, R. A. et al. "Transient convective uplift of an ancient buried landscape." *Nature Geoscience* 4.8 (2011): 562-565.

737   See https://en.wikipedia.org/wiki/File:Globald.png (Accessed on 27 July 2013) The author granted anyone the right to use this work for any purpose, without any conditions.

738   Ceramicola, S., et al. "Anomalous Cenozoic subsidence along the 'passive' continental margin from Ireland to mid-Norway." *Marine and petroleum geology* 22.9 (2005): 1045-1067.

739   "Anomalous Cenozoic subsidence along the 'passive' continental margin from Ireland to mid-Norway" *supra* at p 1062/3

740   "Anomalous Cenozoic subsidence along the 'passive' continental margin from Ireland to mid-Norway" *supra* at p 1063

741   "Anomalous Cenozoic subsidence along the 'passive' continental margin from Ireland to mid-Norway" (See endnote 727) at pp 1056 and 1057

742   Wikipedia describes the Rockall Basin as "*a large (ca. 800 km by 150 km) sedimentary basin that lies to the west of Ireland and the United Kingdom beneath the major deepwater area known as the Rockall Trough*". See http://en.wikipedia.org/wiki/Rockall_Basin (Accessed on 20 September 2014.)

743   The Rockall Basin must not be confused with the Hatton, or Hatton Rockall, Basin. The latter is the 'valley', or 'plain', between the Hatton and Rockall Banks. According to Wikipedia "*The Hatton Basin is a sedimentary basin, located off the west coast of Ireland. It lies between the Hatton and Edoras Banks to the west and the Rockall Bank to the east. The age of the basin fill is uncertain, as its relationship to the Rockall Basin.*" See http://en.wikipedia.org/wiki/Hatton_Basin (Accessed on 20 September 2014)

744   Zhirov, N.. *Atlantis: Atlantology: Basic Problems*. University Press of the Pacific, 2001.

745   At page 301

746   http://en.wikipedia.org/wiki/Lithosphere (Accessed on 8 June 2015)

747   http://en.wikipedia.org/wiki/Earth (Accessed on 8 June 2015)

748   http://en.wikipedia.org/wiki/Earth (Accessed on 8 June 2015)

749   http://en.wikipedia.org/wiki/Lithosphere (Accessed on 8 June 2015)

750   http://goo.gl/zDDQ23 (Accessed on 5 June 2015)

751   Petford, N. et al. "Granite magma formation, transport and emplacement in the Earth's crust." *Nature* 408.6813 (2000): 669-673.

752   Anderson, D. L. and S. D. King. "Driving the Earth machine?." *Science* 346.6214 (2014): 1184-1185.

753   http://en.wikipedia.org/wiki/Magma (Accessed on 8 June 2015)

754   http://en.wikipedia.org/wiki/Asthenosphere (Accessed on 8 June 2015)

755   http://en.wikipedia.org/wiki/Asthenosphere (Accessed on 8 June 2015)

756   http://en.wikipedia.org/wiki/Magma (Accessed on 8 June 2015)

757   Wikipedia defines a pluton as "a body of intrusive igneous rock ... that is crystallized from magma slowly cooling below the surface of the Earth." See https://en.wikipedia.org/wiki/Pluton (Accessed on 17 June 2015)

758   http://en.wikipedia.org/wiki/Magma (Accessed on 8 June 2015)

759   The existence of mantle plumes is not, as yet, universally accepted, however the concept appears to be gaining acceptance.

760   Campbell, . H. "Large igneous provinces and the mantle plume hypothesis." *Elements* 1.5 (2005): 265-269.

761   "A diapir ... is a type of geologic intrusion in which a more mobile and ductily deformable material is forced into brittle overlying rocks. Depending on the tectonic environment, diapirs can range from idealized mushroom-shaped ... structures in regions with low tectonic stress ... to narrow dikes of material that move along tectonically induced fractures in surrounding rock. See http://en.wikipedia.org/wiki/Diapir (Accessed on 8 June 2015) Diapirs are often envisaged as being mushroom shaped.

762   http://en.wikipedia.org/wiki/Mantle_plume (Accessed on 8 June 2015)

763   Obtained from http://www.eurekalert.org/multimedia/pub/91103.php (Accessed on 11 June 2015) created by Mark Richards et al, UC Berkeley with no restrictions on its use.

764   Obtained from http://en.wikipedia.org/wiki/Xiong%27er_Volcanic_Belt (Accessed on 11 June 2015) The author , DBoyd13, licensed it under the creative commons licence.(https://creativecommons.org/licenses/by-sa/3.0/deed.en)

765   http://en.wikipedia.org/wiki/Isostacy (Accessed on 3 April 2018)

766   https://en.wikipedia.org/wiki/Isostatic_depression (Accessed on 23 November 2018)

767   See https://en.wikipedia.org/wiki/Post-glacial_rebound (Accessed on 23 November 2018)

768   http://en.wikipedia.org/wiki/Isostacy (Accessed on 3 April 2018)

769   http://en.wikipedia.org/wiki/Isostacy (Accessed on 3 April 2018)

770   Based upon diagram obtained from http://en.wikipedia.org/wiki/Isostacy (Accessed on 11 June 2015) The author , Gaianauta, licensed it under the creative commons licence. (https://creativecommons.org/licenses/by-sa/4.0/deed.en)

771   http://en.wikipedia.org/wiki/Isostacy (Accessed on 3 April 2018)

772   http://en.wikipedia.org/wiki/Isostacy (Accessed on 12 June 2015)

773   https://en.wikipedia.org/wiki/Sea_level_rise (Accessed on 12 June 2015)

774   http://worldoceanreview.com/en/wor-1/coasts/sea-level-rise/ (Accessed on 12 June 2015)

775   Carminati, E. and C. Doglioni. "North Atlantic geoid high, volcanism and glaciations." *Geophysical Research Letters* 37.3 (2010) and Lambeck, K. "Late Pleistocene and Holocene sea-level change in Greece and south-western Turkey: a separation of eustatic, isostatic and tectonic contributions." *Geophysical Journal International* 122.3 (1995): 1022-1044.

776   These change in stresses could have been sufficient to cause a, relatively speaking, 'instantaneous' elastic deformation of the Earth's surface. See Lambeck, K. "Late Pleistocene and Holocene sea-level change in Greece and south-western Turkey: a separation of eustatic, isostatic and tectonic contributions." *supra* at p. 1024

777   Lambeck, K. "Late Pleistocene and Holocene sea-level change in Greece and south-western Turkey: a separation of eustatic, isostatic and tectonic contributions." *Geophysical Journal International* 122.3 (1995): 1022-1044.

778   Blundell, D. J., and D. A. Waltham. "A possible glacially-forced tectonic mechanism for late Neogene surface uplift and subsidence around the North Atlantic." *Proceedings of the Geologists' Association* 120.2 (2009): 98-107.

779   GSA Today, September 2009, v. 19, no. 9, doi: 10.1130/GSATG50GW.1 , http://en.wikipedia.org/wiki/Post-glacial_rebound (Accessed on 20 September 2014) and http://goo.gl/61mnxQ (at telegraph.co.uk) (Accessed on 20 September 2014)

780   Clark, C. D. et al. "Pattern and timing of retreat of the last British-Irish Ice Sheet." *Quaternary Science Reviews* 44 (2012): 112-146.

781 Clark, C. D. et al. "Pattern and timing of retreat of the last British-Irish Ice Sheet." *Quaternary Science Reviews* 44 (2012): 112-146.

782 http://en.wikipedia.org/wiki/Post-glacial_rebound and http://en.wikipedia.org/wiki/File:Post-glacial_rebound_in_British_Isles.PNG (Accessed on 25 October 2014) The author , Kentynet, licensed it under the creative commons licence. (https://creativecommons.org/licenses/by-sa/3.0/)

783 Sometimes known as a 'magma plume'.

784 Miller, M. S. and T. W. Becker. "Reactivated lithospheric-scale discontinuities localize dynamic uplift of the Moroccan Atlas Mountains." *Geology* 42.1 (2014): 35-38.

785 See also Missenard, Y. et al. "Crustal versus asthenospheric origin of relief of the Atlas Mountains of Morocco." *Journal of Geophysical Research: Solid Earth (1978–2012)* 111.B3 (2006).

786 See further Harris, R. N., and M. K. McNutt. "Heat flow on hot spot swells: Evidence for fluid flow." *Journal of Geophysical Research: Solid Earth (1978–2012)* 112.B3 (2007). and McNutt, M. and A. Bonneville. "A shallow, chemical origin for the Marquesas Swell." *Geochemistry, Geophysics, Geosystems* 1.6 (2000).

787 Approximately 1 600 miles.

788 Jbel Toubkal (also known as Mount Toubkal) is the highest peak with an elevation of 4 167 metres (13 671 feet) in south-western Morocco. See http://en.wikipedia.org/wiki/Atlas_Mountains (Accessed on 3 April 2018)

789 See http://en.wikipedia.org/wiki/High_Atlas (Accessed on 3 April 2018)

790 See http://en.wikipedia.org/wiki/High_Atlas (Accessed on 3 April 2018)

791 See http://en.wikipedia.org/wiki/Middle_Atlas

792 Pascal, C. and O. Olesen. "Are the Norwegian mountains compensated by a mantle thermal anomaly at depth?" *Tectonophysics* 475.1 (2009): 160-168.

793 Weidle, C. and V. Maupin. "An upper-mantle S-wave velocity model for Northern Europe from Love and Rayleigh group velocities." *Geophysical Journal International* 175.3 (2008): 1154-1168.

794 See Helmholtz Centre Potsdam - GFZ German Research Centre for Geosciences. "Magma pancakes beneath Indonesia's Lake Toba: Subsurface sources of mega-eruptions." ScienceDaily. ScienceDaily, 30 October 2014. <www.sciencedaily.com/releases/2014/10/141030142024.htm> and Jaxybulatov, K., et al. "A large magmatic sill complex beneath the Toba caldera." *Science* 346.6209 (2014): 617-619.

795 http://en.wikipedia.org/wiki/Rockall

796 http://en.wikipedia.org/wiki/Hasselwood_Rock (Accessed 3 April 2018)

797 Wikipedia http://en.wikipedia.org/wiki/Iceland_plume (Accessed 3 April 2018)

798 White, R. and D. McKenzie. "Magmatism at rift zones: the generation of volcanic continental margins and flood basalts." *Journal of Geophysical Research: Solid Earth (1978–2012)* 94.B6 (1989): 7685-7729.

799 Hartley, R. A. et al. "Transient convective uplift of an ancient buried landscape." *Nature Geoscience* 4.8 (2011): 562-565.

800 See also Poore, H. R., N. White, and S. Jones. "A Neogene chronology of Iceland plume activity from V-shaped ridges." *Earth and Planetary Science Letters* 283.1 (2009): 1-13.

801 See also Clift, P. D., and J. Turner. "Dynamic support by the Iceland Plume and its effect on the subsidence of the northern Atlantic margins." *Journal of the Geological Society* 152.6 (1995): 935-941.

802 Poore, H., N. White and J. Maclennan. "Ocean circulation and mantle melting controlled by radial flow of hot pulses in the Iceland plume." *Nature Geoscience* 4.8 (2011): 558-561.

803 See also Rudge, J. F. et al. "A plume model of transient diachronous uplift at the Earth's surface." *Earth and Planetary Science Letters* 267.1 (2008): 146-160.

804 Arrowsmith, S. J. et al. "Seismic imaging of a hot upwelling beneath the British Isles." *Geology* 33.5 (2005): 345-348.

805 See also Stoker, M. S., et al. "Neogene evolution of the Atlantic continental margin of NW Europe (Lofoten Islands to SW Ireland): anything but passive." *Petroleum Geology: North West Europe and Global Perspectives: Proceedings of the 6th Conference.* Geological Society of London, 2005., *Clift, P. D. and J. Turner. "Dynamic support by the Iceland Plume and its effect on the subsidence of the northern Atlantic margins." Journal of the Geological Society 152.6 (1995): 935-941.* **and** Champion, M. E., et al. "Quantifying transient mantle convective uplift: An example from the Faroe-Shetland basin." *Tectonics* 27.1 (2008).

806 Davis, M. W. et al. "Crustal structure of the British Isles and its epeirogenic consequences." *Geophysical Journal International* 190.2 (2012): 705-725.

807 See also *Clift, P. D. and J. Turner. "Dynamic support by the Iceland Plume and its effect on the subsidence of the northern Atlantic margins." Journal of the Geological Society 152.6 (1995): 935-941.*

808 Rickers, F., A. Fichtner, and J. Trampert. "The Iceland–Jan Mayen plume system and its impact on mantle dynamics in the North Atlantic region: Evidence from full-waveform inversion." *Earth and Planetary Science Letters* 367 (2013): 39-51.

809 Pfeffer, Julia, *et al.* "Decoding the origins of vertical land motions observed today at coasts." *Geophysical Journal International* 210.1 (2017): 148-165 at p. 149

810 Pfeffer, Julia, *et al.supra* at p. 149

811 Petford, N. et al. "Granite magma formation, transport and emplacement in the Earth's crust." *Nature* 408.6813 (2000): 669-673.

812 Cembrano, J. and L. Lara. "The link between volcanism and tectonics in the southern volcanic zone of the Chilean Andes: A review." *Tectonophysics* 471.1 (2009): 96-113.

813 See also Helmholtz Centre Potsdam - GFZ German Research Centre for Geosciences. "Off-rift volcanoes explained: Crustal unloading." ScienceDaily. ScienceDaily, 23 March 2014. <www.sciencedaily.com/releases/2014/03/140323151958.htm> and Maccaferri, Francesco, et al. "Off-rift volcanism in rift zones determined by crustal unloading." *Nature Geoscience* (2014).

814   Duggen, S. et al. "Flow of Canary mantle plume material through a subcontinental lithospheric corridor beneath Africa to the Mediterranean." *Geology* 37.3 (2009): 283-286.

815   See Blundell, D. J., and D. A. Waltham. "A possible glacially-forced tectonic mechanism for late Neogene surface uplift and subsidence around the North Atlantic." *Proceedings of the Geologists' Association* 120.2 (2009): 98-107.

816   Chaussard, E. and F. Amelung. "Regional controls on magma ascent and storage in volcanic arcs." *Geochemistry, Geophysics, Geosystems* 15.4 (2014): 1407-1418.

817   Dam, G., M. Larsen, and M. Sønderholm. "Sedimentary response to mantle plumes: implications from Paleocene onshore successions, West and East Greenland." *Geology* 26.3 (1998): 207-210.

818   Boldreel, L. O. and M. S. Andersen. "Tertiary development of the Faeroe-Rockall Plateau based on reflection seismic data." *Bulletin of the Geological Society of Denmark* 41.2 (1994): 162-180.

819   Dvorak, J. J. and D. Dzurisin. "Volcano geodesy: The search for magma reservoirs and the formation of eruptive vents." *Reviews of Geophysics* 35.3 (1997): 343-384.

820   See also Field, L., et al. "Magma storage conditions beneath Dabbahu Volcano (Ethiopia) constrained by petrology, seismicity and satellite geodesy." *Bulletin of volcanology* 74.5 (2012): 981-1004.

821   Chaussard, E. and F. Amelung. "Regional controls on magma ascent and storage in volcanic arcs." *Geochemistry, Geophysics, Geosystems* 15.4 (2014): 1407-1418.

822   See also Mitrovica, J. X., and G. A. Milne. "On the origin of late Holocene sea-level highstands within equatorial ocean basins." *Quaternary Science Reviews* 21.20 (2002): 2179-2190.

823   Frederikse, Thomas, Riccardo EM Riva, and Matt A. King. "Ocean Bottom Deformation Due To Present-Day Mass Redistribution and Its Impact on Sea Level Observations." *Geophysical Research Letters* 44.24 (2017)

824   Champion, M. E., et al. "Quantifying transient mantle convective uplift: An example from the Faroe-Shetland basin." *Tectonics* 27.1 (2008).

825   All pictures and graphs are generated from the Digital Terrain Model (DTM) at the European Marine Observation and Data Network at http://portal.emodnet-bathymetry.eu/ (Accessed on 29 June 2015) The DTM with its information layers is made freely available for browsing and downloading (See http://www.emodnet-bathymetry.eu/content/content.asp?menu=0040000_000000) through the Bathymetry Viewing and Download service at http://portal.emodnet-bathymetry.eu/.